Exploring Southern Appalachian Forests

Exploring Southern Appalachian Forests

An Ecological Guide to 30 Great Hikes in the Carolinas, Georgia, Tennessee, and Virginia

Stephanie B. Jeffries

Thomas R. Wentworth

The University of North Carolina Press

Chapel Hill

Publication of this book was supported in part by a generous gift from Cyndy and John O'Hara.

A Southern Gateways Guide

Set in Swift Neue and Gotham Narrow

Manufactured in the United States of America

The text for hike 16, Carlos Campbell Overlook, appeared previously in somewhat different form in Thomas R. Wentworth, "Biological Threats to the Integrity of Southern Appalachian Forests (and What ASB Can Do to Help)," *Southeastern Biology* 56, no. 4 (1 October 2009): 444–49. Reprinted with permission.

The paper in this book meets the guidelines for permanence and durability of the Committee on Production Guidelines for Book Longevity of the Council on Library Resources.

The University of North Carolina Press has been a member of the Green Press Initiative since 2003.

Cover photo by Tom Wentworth.

Library of Congress Cataloging-in-Publication Data
Jeffries, Stephanie B.
Exploring Southern Appalachian forests : an ecological guide to 30 great hikes in the Carolinas, Georgia, Tennessee, and Virginia / Stephanie B. Jeffries and Thomas R. Wentworth.
pages cm.
"A Southern Gateways Guide."
Includes bibliographical references and index.
ISBN 978-1-4696-1979-8 (cloth : alk. paper)
ISBN 978-1-4696-1820-3 (pbk : alk. paper)
ISBN 978-1-4696-1821-0 (ebook) 1. Hiking – Appalachian Region, Southern – Guidebooks. 2. Trails – Appalachian Region, Southern – Guidebooks. 3. Natural history – Appalachian Region, Southern – Guidebooks. 4. Ecology – Appalachian Region, Southern – Guidebooks. 5. Appalachian Region, Southern – Guidebooks. I. Wentworth, Thomas R. II. Title.
GV199.42.A68J45 2014
796.510975 – dc23 2014007166

cloth 18 17 16 15 14 5 4 3 2 1
paper 18 17 16 15 14 5 4 3 2 1

We dedicate this book to our many mentors who led us into the mountains, and to our children, both biological and academic, who follow in our footsteps. One individual in particular has inspired us throughout our journey. J. Dan Pittillo, now retired from Western Carolina University, has captivated a generation of students with his knowledge of the mountains. We dedicate this book especially to Dan, an extraordinary teacher, mentor, and friend.

Contents

Sidebars

Preface and Acknowledgments

Our idea for this project emerged years ago while we were teaching our field course (Forest Ecosystems of the Southern Appalachians) at the Highlands Biological Station. We were hustling along the Fork Mountain Trail out of Ellicott Rock National Wilderness in South Carolina with a dozen students, in the midst of a torrential downpour with accompanying thunder and lightning. The two of us brought up the rear of the party as we dashed for the van, trying to see the trail through sheets of rain. "Steph!" Tom shouted over the roar of the storm and the Chattooga River, "We need to write a book!"

We had been in the middle of a teaching exercise, one in which we ask our students to interpret a forested landscape, before the thunder started and we'd had to make a run for it. Walking into a patch of woods and constructing a story for the forest—what species are present, why that particular group of species is there, what forces have shaped the forest, and what the forest might look like fifty years from now—may seem like a tall order. And that day our exercise had been cut short by the thunderstorm before our students had a chance to really see the forest for themselves. However, during the two weeks of our intensive field course, we teach our students to construct ecological stories for many sites throughout the southern Appalachians.

The book you are holding re-creates for you the experience that we have offered our students. We take you along with us on 30 day hikes in the southern Appalachians, a bit more than 100 miles of hiking if you are able to join us for every hike. Each place we visit has a story, and along the way, we'll share clues that will help you piece that story together. We'll show you how to evaluate the evidence you find, and we'll discuss alternative hypotheses that explain what you see. Always, we ask "why?" Why are some summits forested and others open? Why are these hemlock trees dying? Why do some forests have so many types of trees, while others have few? Why are the trees in this stand all the same size? What did this forest look like 50 or 500 years ago? What will it look like in 10 years? 100?

Along your journey, you'll learn how to see the forest, with the trees.

We firmly believe that anyone can learn to see through an ecological lens. Ecological concepts can be refreshingly intuitive—we have experienced the "aha!" moment more times than we can count, when the different pieces of evidence come together. The challenge lies in knowing what to look for, taking in all the evidence you can find, and weaving it into the unique story for a particular forest.

With time and practice, you will find that all walks in the woods, not just those described in this guidebook, become opportunities to solve mysteries. Your eyes will rove, taking in the canopy trees and understory plants. You'll look for stumps or fire scars and dig into the top layer of soil. You'll note the shape of the landscape and check your compass to determine the direction of the slope. We hope that you will see southern Appalachian forests in an entirely new way.

At the same time, finding new understanding in what we see also makes us realize how little we really know about nature. And the not-knowing puts us in awe of the mystery, resilience, and majesty of these southern mountains.

Acknowledgments

We knew that writing this book would be an enormous undertaking, one where we would need the support and expertise of others. We are grateful for a small army of friends, colleagues, and even strangers who stepped forward to lend a hand.

Our thanks first go to our publisher, UNC Press, particularly our editor, Mark Simpson-Vos, who encouraged us from the earliest stages of this book, and assistant editor Caitlin Bell-Butterfield. We thank our two readers, Mary-Russell Roberson and an anonymous reviewer, whose valuable suggestions added clarity to the text, and Stephanie Ladniak Wenzel, for her editorial guidance. We also very much appreciate Linda Rudd's expert editorial assistance. Financial support for our fieldwork came from the Highlands Biological Foundation, Inc., and the Thompson Writing Program at Duke University.

We had an incredible team of illustrators, all graduates of the North Carolina Botanical Garden's botanical illustration certificate program, create original artwork for our trees section. Emma Skurnick generously offered to advise the talented team of Dale Morgan,

Maryann Roper, and Ruta Schuller. In addition, Carol Ann McCormick at the UNC Herbarium provided technical advice to our illustrators. We thank David Blevins for his support and guidance on writing and photography, and for his time producing very high quality scans of our botanical illustrations. Finally, we thank Scott McGrew for his time and expertise in producing the maps.

When we first considered writing this book, there were hikes that we immediately knew we wanted to share with our readers. In the course of expanding our scope to the entire southern Appalachians, however, we needed help from others who had a favorite trail to share. We received many excellent suggestions via meetings and email, from Mike Brod, Linda Chafin, Pat Cox, Allen De Hart, Leslie Edwards, Laurie Gallagher, Tom Govus, Ed Harrison, Jamie Jones, Josh Kelly, Tom Kenney, Keith Langdon, Jay Leutze, Jim McFalls, Lytton Musselman, Eric Nagy, Howie Neufield, Carl Nordman, Karen Patterson, Bob Peet, J. Dan Pittillo, Joe Pollard, Nancy Raskin, Randy Raskin, Tom Remaley, Dale Suiter, Alan Weakley, Peter White, Becky Wilbur, and Henry Wilbur. In addition, we enjoyed many conversations with fellow hikers on the trail. The yellow surveyor's notebooks we carried everywhere prompted curiosity and resulted in fun conversations and inevitable enthusiasm for our project. We always asked for and received fellow hikers' favorite trail recommendations.

Our technical questions were answered by an incredible team of southeastern plant ecologists, as well as other -ologists, including Jim Costa, Christina Devorshak, Laura Gadd-Robinson, Sean Gallen, Tom Govus, Andy Heckert, Anya Hinkle, Gary Kauffman, J. Dan Pittillo, Milo Pyne, Johnny Randall, Mike Schafale, Chris Ulrey, Alan Weakley, Karl Wegmann, Charlie Williams, Crystal Wilson, Peter White, and Qiu-Yun (Jenny) Xiang. In addition, local trail expertise and history was shared with professionalism and enthusiasm by many state and federal park rangers and foresters, specifically Scott Alexander at Oconee Station State Historic Site, Tim Lee at Jones Gap State Park, and Scott Stegenga at Table Rock State Park. We take responsibility for any errors in the text.

Over the course of covering hundreds of miles of trails, we had many enthusiastic hiking companions who guided us on their favorite hikes, offered us places to stay, carried our flora, or guarded our snacks. Many thanks to Huma Alvarado, Jon Armstrong, Bob Dellinger, Tom Govus, Andrew Jeffries, Simon Jeffries, Stephen Jeffries, Danny Jessup, Jay Leutze, Rob Klein, Patrick McMillan, J. Dan Pittillo, Ed Pivorun, Julie Tuttle, and Alan Weakley for many fun days on the trail.

Several organizations contributed to the success of our project, namely the Great Smoky Mountains Institute at Tremont, Highlands Biological Station, Mountain Lake Biological Station, NatureServe, North Carolina Botanical Garden, North Carolina State University, and the Thompson Writing Program at Duke University.

Three individuals stand out for their support and enthusiasm for the project from its conception. They are David Blevins, Jim Costa, and J. Dan Pittillo.

The inspiration for this book came from many years of coteaching Forest Ecosystems of the Southern Appalachian Mountains at the Highlands Biological Station in odd years, starting in 1997 to the most recent offering in 2013. We have learned from every instructor who has contributed to the course: Bob Peet, J. Dan Pittillo, Julie Tuttle, Alan Weakley, and Peter White. In addition, we thank the many students we have learned from while teaching the course. Their enthusiasm during what is always a demanding and intensive two-week field experience inspired us to reach beyond the class to write this book.

Steph personally thanks three NC State University mentors: Lee Allen, Barry Goldfarb, and Tom Wentworth, for supporting her through her often-unconventional career decisions. Her weekly writing group of Gretchen Case, Seth Dowland, and Keith Wilhite helped move the project forward, and her many students at the North Carolina Botanical Garden kept her enthusiasm for the project high. She wishes that she could have shared these trails with her lifelong friend, fellow scientist, and outdoorswoman Suzie Seemann. She thanks Ann Camden for her unwavering encouragement from the very beginning all the way through to the end. And support from her husband, Andrew, and her running community, Runnerpeeps, kept her on a mostly even keel on the long road to publication.

Tom personally thanks his colleagues who first introduced him to the southern Appalachian Mountains. These individuals, all associated with Highlands Biological Station, are Dick Bruce, John DeLapp, and Jim Hardin. He also thanks Bob Dellinger, Karen Patterson, and

J. Dan Pittillo for many fine days of botanizing on mountain trails. Over a quarter-century of field experience with the hundreds of volunteers in the Carolina Vegetation Survey added greatly to Tom's appreciation of natural communities in the Carolinas, and he is especially grateful for the mentorship of colleagues Bob Peet, Mike Schafale, Alan Weakley, and Peter White. Tom also appreciates the support and intellectual environment provided by Margo Daub and Tom's other colleagues in the Department of Plant and Microbial Biology at North Carolina State University.

Finally, parts of this project have been all-consuming and at times have pulled us away from the ones we love the most. Our respective families' support has meant the world to us and made this endeavor possible. Tom thanks his wife, Linda Rudd, and Steph thanks her husband, Andrew Jeffries, and her sons, Stephen and Simon.

Exploring Southern Appalachian Forests

Introduction

Welcome to what we hope will be a new way for you to appreciate the natural communities of the southern Appalachian Mountains! For seasoned hikers and neophytes alike, we offer a set of guided hikes that will open new doors to understanding the forests, shrublands, grasslands, and rocky areas you hike through. You'll learn to recognize the different natural communities and their component plant species. Beyond this, however, we'll take you "behind the scenes" to discover the inner workings of forests, from plant adaptations to the physical environment to the responses of plants to novel challenges associated with climate change and invasions of exotic pests and diseases. We'll explore the deep evolutionary and biogeographic history of these forests and how the present can only be fully appreciated in the context of the past. Master environmental gradients associated with latitude, elevation, topography, and geology shape the vegetation of our southern mountains, and we'll show you how to interpret these gradients as you travel across the landscape. We'll also consider the everyday processes that make our forests such dynamic, ever-changing entities, including history, disturbance, and recovery. We'll help you appreciate the myriad problems that land managers and conservationists must address in their quest to preserve the biological diversity, productivity, and services provided by natural ecosystems. And finally, we'll pause with you to appreciate the natural beauty and grandeur that make the southern Appalachians one of the most wondrous places on earth.

Readers can approach this guide in a variety of ways, depending on their specific interests. Some will begin by browsing the different hikes, selecting one or more that are appealing because of their setting or destination. Others will begin with a perusal of the natural communities, which we've organized in a single section. In this way we avoid repeatedly describing a given natural community in the hikes where it occurs. After choosing a natural community of interest, the reader can then select one or more hikes that access that community.

Others may begin by perusing the locator map, selecting a few hikes in close proximity to their home or weekend destination. If you choose this approach, look through those hike chapters to decide which hike is the best fit for your needs. You might then wonder what a particular natural community is like, so you can turn to the natural communities section to better understand what makes it unique.

Finally, we hope that all hikers will find our original line drawings of southern Appalachian plant species of special interest. We selected the most commonly encountered woody plants, and these are portrayed by three talented botanical illustrators in the section "Common Trees and Shrubs of the Southern Appalachians." If you learn to recognize these species, among the many hundreds that occur throughout the region, you will be well on your way to being "at home" in a southern Appalachian forest.

As you read this book, you are likely to encounter new and unfamiliar information or terminology. Although we have tried to avoid unnecessary jargon and pedantic ramblings, we do expect that you may be pushed beyond your comfort zone as you begin to use this book. If we didn't lead you into unfamiliar territory, both in the mountains and intellectually, we'd be failing in our goal to help you read the landscape. However, we aim to make you comfortable in new intellectual terrain. We include explanations of ecological phenomena, discussion of new terms and concepts, and a glossary of ecological terms that are fundamental to understanding and communication about ecology.

Another element in this book that may push you out of familiar territory is our use of Latin names for plants. Why can't we just use common English names? There are several reasons. First, some common names are applied to more than one species. Second, many species have multiple common names. What we call striped maple (*Acer pensylvanicum*) in this book is also known variously as moosewood, whistle-wood, and goosefoot maple. Finally, some common names are misleading. What we call tulip-tree (*Liriodendron tulipifera*) in this book is also known as tulip poplar and yellow-poplar. However, tulip-tree is not a tulip and is only distantly related to poplars.

Latin names, although unfamiliar and sometimes challenging, are essential for plant nomenclature and scientific communication. In theory, each species of plant has a unique Latin name, and that name is recognizable to scientists and laypersons alike across the world. Latin names are two words, Linnaean binomials, like the familiar *Homo sapiens*, the binomial name for our species. The first part of the name is the genus, which always has its first letter capitalized, and the second part is the specific epithet, which is never capitalized, even when derived from a proper noun. Latin names should always be either italicized, as in this book, or underlined.

As new information becomes available to plant systematists and taxonomists (the scientists who worry about the relationships among plants and their nomenclature), the widely agreed-upon Latin names may change. Thus it is important to use an up-to-date regional reference for nomenclature. Our choice for this book is Alan S. Weakley's *Flora of the Southern and Mid-Atlantic States*. As for common names, we have simply chosen familiar ones we like and have used them consistently throughout the book.

Hike Information

For each hike in our book, you'll find the following information:

Highlights: A short description of the hike and why you might like to go.

Natural Communities: A list of the main community types on the hike, which you can then read about in the natural communities section.

Elevation: Important for weather considerations as well as the kinds of natural communities encountered.

Distance: Also denotes a loop, point-to-point, or round-trip if it is an out-and-back hike.

Difficulty: A descriptive assessment noting the hike's difficulty but also including information about what makes it so—such as "a moderate climb," "rocky," or "level walking."

Directions: Step-by-step directions from major crossroads to get you to the trailhead. Also tells you what facilities may be available and which, if any, of the facilities are accessible for those with physical disabilities. On that note, although this is a hiking guide, there are several hikes that are accessible or partially accessible. These are 3

(Brasstown Bald), 13 (Hooper Bald), 16 (Carlos Campbell Overlook), 21 (Wiseman's View), and 23 (Roan Gardens).

GPS: Trailhead coordinates (and other important coordinates as needed) are included for each hike.

Information: Contact information for each place. It's always a good idea to call a park or preserve you plan to visit, to make sure there are no closures.

Maps: Maps are included primarily to get you to the trailhead and give you a rough idea of the route. All the trails featured in this book are well-established and easy to follow. Using the map and hike description, you should have no trouble finding your way around on any trail. If you want to hike additional trails nearby or take some of the spur trails or extensions, you will need to get a topographic map, a compass, possibly a GPS unit, and one of the many excellent trail guides available for this region.

Why Are We Here? This is a question we always address as soon as we stop at any new site during our field course. There are literally thousands of trails in the southern Appalachians. Why did we choose this one? What makes it special? What natural communities will we encounter? What are the most interesting ecological lessons?

The Hike: This is the description of each hike, including basic directional information as well as the ecological story for the hike. Each hike also includes a sidebar with in-depth coverage of a specific topic of interest.

Safety and Preparation

All the hikes in this book are on well-marked, established trails. The maps in this book will get you to the trailhead, but they are no substitute for a compass and a topographic map that you can find online, in a local bookstore, or at the visitor center or trailhead. You should always know or have a reference for your route. In addition, someone not in your hiking party should know where you are and when you expect to return. You cannot be certain of cell phone service or GPS reception in mountainous terrain.

Always be ready for changes in weather. The weather in the mountains can turn from wonderful to atrocious in a very short time span. Our fashion advisory is that dressing in layers is your best strategy for being prepared for weather contingencies. Synthetic fabrics next to the skin are far more comfortable than cotton in anything other than ideal conditions. Leave those heavy jeans and T-shirts back at camp for wearing when you lounge around the campfire! Essentials include raingear (which can be as simple as a plastic rain poncho), a warm layer, and a hat. On the most recent iteration of the Forest Ecosystems class, the Highlands, NC, area received more than 14 inches of rain in the first 6 days. And several years ago, a windy, rainy 50°F day in July on Roan Mountain made the down vest and fleece hat that were stuffed into a self-seal bag and carried everywhere an investment that finally paid off. They take up very little space in your pack but are well worth their weight.

While this section is hardly exhaustive, hikers should be aware of common hazards that they may encounter on their outings. You will need to carry a first aid kit with a manual of common symptoms and treatments, basic over-the-counter medications such as pain relievers and antihistamines (for allergic reactions), and plenty of bandages for minor scrapes and cuts. Insect bites and stings, allergic reactions to plants, venomous snakes, hyperthermia, hypothermia, lightning strikes, and falls are all inherent risks for any outdoorsperson, but most of these you can avoid with proper preparation. We feel that the most notable high-risk hazards are hypothermia, lightning strikes, and falls, so we'll elaborate a bit on each of those.

Staying dry, warm, well-nourished, and well-hydrated in the mountains is more than just a matter of comfort. Hikers should be aware of the signs and treatment of hypothermia, a decline in body temperature below the normal range of 98–100°F. If first aid measures fail to reverse the symptoms quickly, you should seek immediate medical attention. If a thunderstorm threatens, the best response is to seek shelter in a building or vehicle. When this is not feasible, you can still minimize the risk of being struck by lightning. Avoid open areas and exposed portions of the landscape (peaks, hilltops, ridges). Never stand near or under isolated tall objects, such as trees or power poles. The safest places outdoors are in topographically protected areas (valleys or ravines), away from the tallest trees and water. Avoid sheltering under rock overhangs or in other situations where you could become part of the shortest path of lightning to ground. Lastly, use caution when you are in wet, high, or exposed places. Each year many people are seriously hurt or killed when they climb around waterfalls, something we strongly advise against. Granitic domes can be especially deceptive, because you can descend over the rounded surface without noticing increasing steepness. Pay attention to your surroundings and watch the edges!

Likewise, be prepared to stay out on the trail longer than you expect. We tell our students to always bring plenty of water and food. This advice was echoed by a former student who advised future students, "Always bring food. Tom and Steph will promise that you'll be back by lunchtime, but they'll get so entranced with the wonders of the plant kingdom that you will run late—sometimes very, very late. It's best to just carry food with you and roll with it, and soon, you'll be caught up in their excitement." In addition, we always carry a flashlight or headlamp, in case we are out later than expected. The bottom line is that you need to be responsible for your own safety. Being prepared and aware of your surroundings and having a contingency plan will get you through most situations you will encounter in the field.

Take More Than Photographs

"Take only photographs; leave only footprints." This "Leave No Trace" ethic allows others who follow in your footsteps to also enjoy spending time in nature without reminders of the built world. We suggest that you take an extra step and always carry a bag to carry out any trash you might find on the trail. Leave the trail better than when you found it, a small token of your appreciation for your experience.

Natural and Human History

A fundamental question for ecologists is "Why are there so many species on earth?" and this is a good starting point for understanding the forests of the southern Appalachians. Great Smoky Mountains National Park, for example, has more tree species than any other national park in the United States. Why? To answer this question and others like it, we must carefully examine the natural and human histories of the region.

The southern Appalachians are an ancient landscape, with some of the oldest rocks in the world, thought to be 1 billion years old. The mountains that you see today were formed and uplifted when the ancestral continental plates of North America and Africa collided to form the Pangaea supercontinent, separated, and collided again repeatedly, starting more than 250 million years ago. Each successive rifting and collision pushed the Appalachians, which stretched from present-day Alabama to Norway, still higher. What we see today are the worn-down, forest-cloaked peaks of a once-mighty range—one that topped the Rockies in elevation and grandeur.

These mountains are old, and so are their species. Unlike the northern reaches and eastern coastal plain of North America, this landscape has never been under ice or water. The thick ice sheets that covered the northern portions of North America during repeated glacial events never reached the southern mountains. Rising seas during interglacial periods drowned coastlines to the east but never reached the foothills. Since terrestrial life-forms first emerged on the continental margins, they have found a home in the southern Appalachians. Here, life specialized, evolved, and took on new forms as the environment demanded, through the process of natural selection. One ancestral species of *Trillium* begat others, many uniquely adapted to their own river basins, as its seeds were laboriously dispersed by ants.

As life flourished in the southern mountains over millions of years, it diversified, responding to changes in climate. Glaciers, until only very recently (just 18,000 years ago), crept southward, though never into the southern mountains, and the evergreen trees now restricted to the highest peaks extended their ranges into cool and moist valleys, leaving the highest peaks with open, alpine meadows above timberline. Warming trends sent spruce and fir ascending the highest peaks and brought an influx of tropical and subtropical species from the south into the lower elevations. The north-south orientation of the southern Appalachians enabled species to migrate along their spine, always finding climates that best suited them. During warmer periods, the southern Appalachians became isolated climatically from the north, allowing speciation to occur. These mountains thus became an ecological crossroads, a meeting place for species from north, south, and central North America and beyond. Many of these species found suitable habitat niches and stayed.

What evidence do we have of these species' migrations? Researchers Paul and Hazel Delcourt from the University of Tennessee–Knoxville took soil cores of sediment from bogs all over the southern mountains to better understand species composition in past climates. Because organic matter doesn't decompose in submerged, low-oxygen bog environments, the cores provided an organic, layered record, including pollen grains that can be dated and identified to genus. Widespread pollen from the spruce genus (*Picea*) suggests that spruce forest once covered most of the southern mountains during a period of cooler climate. This finding also explains why we might find Fraser fir (*Abies fraseri*) on the tops of most mountains taller than 5,000 feet, but not all of them. Warmer temperatures in the past might have forced this species off some mountaintops, while on others it found refuge in north-facing microclimates, later expanding its coverage as temperatures cooled. Because of the distance between the high mountain peaks, Fraser fir has not returned to all of those from which it was extirpated. Thus, some peaks in North Carolina, Tennessee, and Virginia that are tall enough to support Fraser fir today lack this species altogether. Latitudinal gradients also play a role in determining the distribution of spruce-fir forests, with none of this forest type in either Georgia or South Carolina, even though conditions on some of the higher peaks are conducive to its growth. What we see on our hikes today thus holds a record of the past.

Today's mountain climate is important in fostering biodiversity. Elevation is perhaps the easiest way to predict temperature, with approximately 3°F of mean annual temperature lost for every thousand feet gained. In addition to a range of temperatures at different elevations, the southern Appalachians include some of the wettest and driest localities on the East Coast. The wettest areas skirt the borders of North Carolina, Georgia, and South Carolina along the Blue Ridge Escarpment. Here, weather systems approach from the south or west; the warm, moist air rises over the mountains; and precipitation falls over the area in amounts that can total more than 100 inches per year, making this area essentially a temperate rain forest. On the flip side, the area around Asheville, NC, is a basin surrounded by this same escarpment to the south as well as the Great Smoky Mountains to the west, so its annual rainfall is less than half of that in the escarpment gorges. By the time weather systems arrive in the Asheville basin, they have dumped their precipitation on the higher surrounding mountains, creating a rain shadow effect.

Topography can help determine the natural community and thus the combination of species found at a particular site. Bowl-shaped landforms, sometimes called coves, contain deeper soil and are sheltered from extremes in weather, high winds, blazing sun, and wind-driven ice and snow. Slopes, depending on their compass aspect and steepness, can receive varying degrees of solar radiation, and at different times of the day. Exposed ridgelines often have thin, rocky, and drought-prone soils.

On a smaller scale, microtopography helps dictate availability of soil moisture as well. Several rare fern species feel most at home in the constant spray behind waterfalls. Tiny jewels of wildflowers, together called spring ephemerals, complete their life cycles at the feet of tall hardwoods before these trees leaf out in early spring. Some species that are poor competitors live on the margins; by surviving challenging habitats like dry, open rock faces, they live where other species cannot.

Every species has a unique niche, a place where it best grows and thrives. The many climates and microclimates, and habitats and microhabitats, offer opportunities for specialists and generalists alike to thrive in our southern mountains.

Although moisture and temperature are excellent predictors and properties of natural communities and habitat types, they are by no means the only important factors at work. What lies beneath the surface, in terms of mineralogy, plays an additional and important role. Rocks erode, becoming the mineral parent material of the soil. Soils, in turn, are required by most plants for their supplies of nutrients and water, as well as anchorage to the land. Depending on how rocks were formed and when, soils can vary greatly in terms of pH and nutrient availability for plants. For example, much of the Blue Ridge Mountains consists of granite, which has been transformed (or metamorphosed) by heat and pressure into gneiss. Granites and gneisses weather into soils that tend to be acidic, with pH values as low as 3 or 4, similar to that of lemon juice. Acidic soils do not lack nutrients, but essential minerals such as calcium, magnesium, and potassium are less readily available to plants. The acidity also means that other essential elements, such as phosphorus, remain largely in insoluble form, while aluminum, which is toxic to most plants in large quantities, becomes more available. On the flip side, places such as the Amphibolite Mountains in northwestern North Carolina have "sweeter" soils (higher pH, though still slightly on the acid side of neutral). Nutrients like calcium, magnesium, potassium, and phosphorus are more available for plants in the Amphibolites, while aluminum availability and toxicity are reduced. As you accompany us on our hikes, you will learn to look for clues in the vegetation that can help you decide whether a site is acidic or "rich" (closer to neutrality, or "circumneutral," and higher in available soil nutrients) without the need for a soil test. Because plants are often particular about local conditions, they are excellent indicators of such conditions, including soil properties. For example, acid-loving plants include most species in the Ericaceae family, while the presence of basswood (*Tilia americana*), cucumber magnolia (*Magnolia acuminata*), and white ash (*Fraxinus americana*) together is pretty good evidence of a rich, circumneutral soil. Other species, such as eastern hemlock (*Tsuga canadensis*), tulip-tree (*Liriodendron tulipifera*), and red maple (*Acer rubrum*), are more broadly tolerant of a range of conditions and are thus less useful indicators of site quality.

The physical environment is thus critically important in determining the distributions of plant species and natural communities, but it does not fully explain the biological diversity of the southern Appalachians. Underappreciated factors that have wide-ranging implications for the distribution of species and natural communities are spatial and temporal constraints. Species cannot occur in suitable habitats to which they do not have access. As noted earlier, we suspect that some high-elevation peaks in the southern Appalachians lack Fraser fir and/or red spruce (*Picea rubens*) because these species have been unable to recolonize sites where they were eliminated during times of warmer climate. Past history leaves its indelible imprint on natural communities, as seen in the remarkable similarity of eastern North American and eastern Asian floras, a legacy of connections between these areas that existed tens of millions of years ago.

Disturbance can sweep away an existing natural community, leaving an opening for different species and vegetation to regenerate. Disturbance can take the form of storm damage resulting from a tornado, hurricane, or winter storm that destroys a swath of existing forest. Fires, whether caused by lightning or set intentionally by humans, can damage or destroy an existing natural community and set the stage for succession. Some species, such as tulip-tree, cannot tolerate shaded conditions for growth, so their presence in a forest, particularly if their stems are of the same age, can tell us something about a site's past. Likewise, many pines, such as pitch (*Pinus rigida*) and Table Mountain pine (*P. pungens*), are especially well-adapted to fire. In both cases, these species get a foothold only when a gap in the canopy is opened by a disturbance—either naturally (lightning-ignited fire or storms) or human-initiated (logging or clearing for agriculture)—that removes the vegetation. In addition, some of these shade-intolerant species are not especially long-lived, so their presence can be a good indicator of how recently a site was disturbed. When these pioneers fall, they create new openings in the forest. Disturbance and the time elapsed since the last disturbance both add additional layers to the complexity and diversity of natural communities in the southern mountains.

Although natural disturbances, such as windstorms and fire, affect communities in important ways, human disturbances are even more pervasive. There are still places, such as Joyce Kilmer Memorial Forest and parts of the Smokies, where you can walk into a cathedral-like forest of giant trees, but few still exist. The human reach has been long, it has been persistent, and it has taken on many forms. Before settlement of the region by Europeans, Native Americans used fire to clear the forest understory to promote hunting, and small tracts were cleared for agriculture. European settlers, arriving early in the 18th century, continued small-scale landscape transformation as they created homesteads and farms in many areas that still had pristine forests. Forest resources were abundant, and the initial impact was small, mostly caused by small-scale logging, subsistence agriculture, and grazing by domestic animals. A gold rush in the early part of the 1800s brought more settlers to the region, and the last of the native people were driven out, most notably and tragically on the Trail of Tears for the Cherokee during the winter of 1837–38.

The small patches of disturbance to the land caused by subsistence communities were replaced by large-scale clearing with the arrival of railroads in the mid-1800s. Not only was much of the pristine forest cut, but the logging practices of the time left lasting scars on the land that have been slow to heal. Cable logging dragged huge trees across the land, ripping up the soil; fires were set to the leftover logging debris and often raged up the sides of mountains, destroying the forest and burning into its soil. Splash dams (temporary wooden dams built by loggers) were filled with water and floating logs, then dynamited to create a slurry of water and logs that surged down a valley to the nearest rail landing. The rainstorms that followed caused landslides of debris that laid waste to the steep slopes. Scars of these larger-scale disturbances can still be seen throughout the mountains today.

Following the Civil War and into the turn of the 20th century, it became clear that conservation was needed in response to the rapid depletion of forests in the mountains. The conservation movement began as a concern for protecting the remaining timber resources for the future. Over time, however, preservation for the future included the land itself. In 1934, backed by pennies donated by the children of North Carolina and Tennessee, federal money, and donations by philanthropist John D. Rockefeller, Great Smoky Mountains National Park was dedicated. Only 20 percent of the preserved park land included pristine, never-cut forests, and the national park displaced more than 1,200 families that lived within its boundaries. Current park management strikes a fine balance between preservation of natural and human history, respecting those whose ancestors were evicted by preserving some of their cultural heritage in areas such as Cades Cove.

Today, as the human population grows, human disturbance to the mountains continues apace. Disruption of natural disturbance cycles, like fire suppression, negatively affects wilderness areas that used to have a regular fire cycle. Air pollution, much of it produced by motor vehicles and industrial activities to the north and west, causes stressful conditions for many plant communities in the southern mountains, particularly in the higher elevations. Species introduced from around the world have invaded and established a foothold, transforming large swaths of forest in a few short years by wiping out dominant trees in the canopy. Finally, the elephant in the room, rapid global climate change, threatens natural communities in the mountains more than in other places, though we are not yet certain of the future severity and extent of this threat.

The hikes in this book and elsewhere in the southern Appalachians take you on trails that weave through a remarkably diverse and complex system of natural communities. These communities are shaped by climate, geology and soils, topography, biogeography, natural and human history, and time. Each of these factors contributes to the amazing diversity you see around every bend in the trail. Nature is resilient, and forests now stand where there were once pastures. Clear-cuts have healed. Species persist. Others disappear. Erosion continues. Continents drift. And time passes. All processes leave clues in the landscape that you can learn to recognize. The ecological story that emerges is one that is fascinating, complex, and still unfolding.

Natural Communities of the Southern Appalachians

Why Classify Natural Communities?

When seeing the forests while hiking the southern Appalachian Mountains or elsewhere, you've no doubt noticed different kinds of natural communities, assemblages of species that occur predictably in the landscape in particular habitats. For example, you may have observed that some forests have mostly evergreen trees, others are all hardwoods, and some areas lack trees entirely. Or you may have noticed rhododendrons hugging the banks of small streams, or pine trees perched precariously on rocky ridges. Recognizing and naming different kinds of communities will help you understand the astonishing complexity of the vegetation you see, as you identify associations among species and their corresponding environmental conditions. Calling a forest a "spruce-fir forest" or a "cove forest" conjures up images of the forest's appearance, its physical environment, and perhaps some of the plants and animals found there. Thus, classifying communities helps us discuss them usefully – with important applications for conservation and management. It would be difficult to use this book if it did not have a classification of natural communities as its "backbone."

We adopted a widely used national classification of ecological systems developed by NatureServe (a nonprofit conservation organization headquartered in Arlington, VA) that works well for our purposes of introducing you to southern Appalachian forests and related natural communities. This particular system is broad enough to have recognizable distinctions among community types, with enough detail to appreciate the subtleties. Here we provide a capsule description of each of the 12 broad ecological systems (which we will call natural communities) referenced in the different hikes presented in this book. Reading these descriptions is a good way to learn about the natural communities of the southern Appalachians, just as a review of our species sketches is a good way to learn about some of the common woody plant species of the region. You may then return to these community descriptions as you encounter different kinds of forests on your hikes.

Challenges with Classifying Communities

Unfortunately, natural communities are far more difficult to classify than plants or animals because communities overlap broadly as the environment varies from place to place. In natural communities, we don't find the kinds of genetic, reproductive, and morphological barriers that separate, for example, domestic dogs from domestic cats or red spruce (*Picea rubens*) from Fraser fir (*Abies fraseri*). You've no doubt noticed that the species composition and other characteristics of forests often change gradually as you move from one place to another along a hiking trail or roadway. In some sense, natural communities defy classification, and you may be surprised to know that many scientists do not believe that concepts of natural communities correspond well to real entities in the landscape. However, most scientists agree that classification is useful. As a result, they have invested time and effort to develop a comprehensive national classification for natural communities of the United States. To learn more about the scientific debate over the reality of natural communities, please refer to the sidebar in hike 15 (Gregory Bald), "Community Concepts in Conflict."

A word to the wise: after learning about different kinds of forest communities, do not be surprised if you find yourself scratching your head when standing in a real patch of forest and trying to decide which natural community it represents. Keep in mind that much of the landscape consists of forests that are transitional from one "type" to another. Disturbance can also alter forests in ways that make them difficult to classify. If you keep these things in mind, you won't become frustrated, and you'll enjoy the greater perspective that having a working classification lends to your appreciation of our southern Appalachian forests.

How to Identify Natural Communities

Deciding which natural community you're observing (or standing in) requires that you consider the site carefully. First, what is the general appearance and structure of the vegetation? Are you in a forest, woodland, shrubland, or grassland? Or are you in a more sparsely vegetated area with exposed soil or rock? Although most of the natural communities featured in this book are forests, some are woodlands (montane pine forest and woodland), shrublands (shrub bald), grasslands (grass bald), or more sparsely vegetated areas (granitic dome, seepage wetland, and rocky summit).

Next, try to identify the dominant species—those with the greatest ground cover and biomass. If, for example, you are in a stand dominated by red spruce, Fraser fir, or some mixture of the two, you are most likely in a spruce-fir forest. Some natural communities have multiple dominants. As an example, northern hardwood forest may consist of a mixture of American beech (*Fagus grandifolia*), yellow birch (*Betula alleghaniensis*), sugar maple (*Acer saccharum*), and yellow buckeye (*Aesculus flava*). This might lead you to think that all northern hardwood forests must have all four of these species, but many good examples lack one or more of these species. High-elevation boulderfields may be nearly 100% yellow birch, and high-elevation "gaps" (low places along ridgelines) may be dominated by American beech. However, these are all subtypes of northern hardwood forest!

Now, take stock of the local environment. Elevation dictates climate in the mountains. You will rarely find spruce-fir forest below 4,500 ft because conditions are simply too warm and dry for the dominant species to thrive. You are unlikely to find a low-elevation pine forest above 4,000 ft. At low to mid-elevations, topographic position and aspect are also important factors to consider. Topographic position simply refers to where you are relative to the landscape. Are you in a protected site such as a cove or ravine, on an open slope that is neither protected nor exposed, or on a ridge or summit? Protected sites typically have deeper soils and are sheltered from the drying effects of sun and wind, so they are cooler and moister than surrounding landforms. Ridges and summits typically have shallow soils and may be hotter and drier than their more sheltered counterparts. Aspect refers to the compass bearing of a slope, or the direction it faces. In the Northern Hemisphere, southerly aspects (southeast- to southwest-facing slopes) receive much more solar radiation than do northerly aspects (northwest- to northeast-facing slopes). Imagine hiking along a trail that stays at 3,000 ft, but one that takes you from a sheltered, north-facing cove, through open slopes that face east, to a steep ridge that faces south. Even though you haven't gained or lost elevation, you'll hike from a cove forest through an oak forest to a montane pine forest and woodland!

Finally, did we mention that forests often vary continuously in their composition as the environment changes? Cove forests can intergrade gradually into northern hardwood forests that intergrade into spruce-fir forests with increasing elevation. It is quite acceptable, and a testament to your keen observation skills, to say that you are in the transition between a northern hardwood forest and a spruce-fir forest.

Some species have broad ecological tolerances, and these are found in many communities. Red maple (*Acer rubrum*) is notorious for being found just about everywhere, so it is rarely used as a "diagnostic" species for community classification. But even species that are helpful in identifying certain natural communities may be found as components of other natural communities. American beech is a good example. Its presence (especially when dominant) can be a good indication of a northern hardwood forest, but American beech also occurs in cove forests and elsewhere.

Past history of a given site can also influence the community type, making it difficult to classify. Even-aged forest stands of single dominants, especially aggressive invaders of abandoned fields like tuliptree (*Liriodendron tulipifera*) and white pine (*Pinus strobus*), may tell us more about past history of a site (the fact that it was an open, agricultural field 50 years ago or that the site was logged, so the trees there today all germinated at once) than what that site might eventually support under more natural conditions. On the flip side, some forests thrive on disturbance. A hot fire on a steep, south-facing slope can stimulate regeneration of a low-elevation pine forest. However, if fire is suppressed for many decades, the pine forest will gradually decline as the pines die from attacks of southern pine beetle, lightning strikes, wind-throw, and old age. Lacking fire to open the canopy and prepare a good seedbed for pine regeneration, such sites will revert to forests of drought-tolerant hardwoods, like scarlet oak (*Quercus coc-*

cinea), black gum (*Nyssa sylvatica*), and sourwood (*Oxydendrum arboreum*). (Please refer to the sidebar "Fire on the Mountain" in hike 21 [Babel Tower and Wiseman's View] for more information on this topic.)

Please read our descriptions of the southern Appalachian communities covered in this book and try your hand at identifying some of them on your next hike. Our guided hikes will help you learn the various communities quickly. The main point is to enjoy your hikes as you learn more about the fascinating natural communities that surround you!

The Ecological Systems

The table is a key for the formal names used by NatureServe and the briefer community names we chose for this book.

NatureServe Ecological System	Community Name
Forest and Woodland	
Central and Southern Appalachian Montane Oak Forest	Montane Oak Forest
Central and Southern Appalachian Spruce-Fir Forest	Spruce-Fir Forest
Southern and Central Appalachian Cove Forest	Cove Forest
Southern Appalachian Low-elevation Pine Forest	Low-elevation Pine Forest
Southern Appalachian Montane Pine Forest and Woodland	Montane Pine Forest and Woodland
Southern Appalachian Northern Hardwood Forest	Northern Hardwood Forest
Southern Appalachian Oak Forest	Oak Forest
Upland Grassland and Herbaceous	
Southern Appalachian Grass and Shrub Bald	(subdivided for our purposes)
Grass Bald	Grass Bald
Shrub Bald	Shrub Bald
Herbaceous Wetland	
Southern Appalachian Seepage Wetland	Seepage Wetland
Sparsely Vegetated	
Southern Appalachian Granitic Dome	Granitic Dome
Southern Appalachian Rocky Summit	Rocky Summit

Natural Community Overviews

Montane Oak Forest

Oak-dominated forests are by far the most common natural communities in the southern Appalachian Mountains over a wide range of elevations, so it should be no surprise that hardy oaks can often be found in mid- to high-elevation stands (3,000 ft and as high as 5,500 ft). These stands are typically dominated by red oak (*Quercus rubra*) and, less frequently, white oak (*Q. alba*), or mixtures of the two. Identifying the dominant oaks is important, so please refer to the sidebar in hike 7 (Fork Mountain Trail), "Southern Appalachian Oaks," as well as the illustrated Common Trees and Shrubs of the Southern Appalachians.

Montane oak forests are usually found on exposed upper slopes and ridges, where acidic, shallow soils provide relatively few nutrients, unreliable moisture, and poor physical support. Unlike their lower-elevation counterparts, these oak-dominated forests are usually short in stature, typically only 20 to 50 ft in height. The trees themselves are often contorted because of damage sustained from high winds, particularly in the winter months, when ice and snow loading can break branches. Because they resemble apple trees in an old, untended orchard, the stands are often called oak orchards.

Although red and white oaks are the canopy dominants, additional trees occur in these stands, including black cherry (*Prunus serotina*), sweet birch (*Betula lenta*), and yellow birch (*B. alleghaniensis*). American chestnut (*Castanea dentata*) was formerly an important codominant, as evidenced by the abundance of chestnut woody debris and active sprouts seen in many stands. Occasional chestnut sprouts may achieve flowering and fruiting size before succumbing to the chestnut blight. The understory is typically woody, dominated by a diverse assemblage of small trees and shrubs, such as striped maple (*Acer pensylvanicum*), smooth serviceberry (*Amelanchier laevis*), witch hazel (*Hamamelis virgi-*

niana), and flame azalea (*Rhododendron calendulaceum*). Herbaceous cover and diversity are generally low, although grasses, sedges, and ferns may form dense swards in areas where shrub competition is lower. Occasional stands may have a well-developed and species-rich herbaceous layer more reminiscent of cove forests. At lower elevations, montane oak forests typically grade into cove forests, oak forests, or montane pine forests and woodlands. At higher elevations and in more protected sites, northern hardwood forests replace the montane oak forests. Montane oak forests may also grade directly into spruce-fir forests on south-facing slopes.

Scientists believe that montane oak forests burned occasionally in presettlement and early settlement times, perhaps on a 40- to 60-year cycle. Infrequent, low-intensity fires may have favored oak regeneration. Since effective fire suppression has been in place for nearly a century, oak reproduction is declining in many stands, and fire-intolerant species, like sugar maple (*Acer saccharum*), are increasing. Please refer to the sidebar in hike 21 (Babel Tower and Wiseman's View), "Fire on the Mountain," for more information about the effects of fire on natural communities.

Logging played a lesser role in montane oak forests than in the lower-elevation oak forests, but both younger, even-aged and older, all-aged forests may be encountered. Chestnut blight remains a threat, and gypsy moth, although still not established in the southern Appalachians, is a potential threat. Like all high-elevation forest communities in the southern Appalachians, montane oak forests will be affected by climate change.

Hikes with fine examples of montane oak forests include 1 (Blood Mountain), 3 (Brasstown Bald), 4 (Tennessee Rock Trail), 12 (Whiteside Mountain), 15 (Gregory Bald), 22 (Table Rock), and 27 (Mount Jefferson).

Spruce-Fir Forest

Above 5,000 ft (sometimes as low as 4,500 ft), forests of the southern Appalachians are dominated by evergreen red spruce (*Picea rubens*) and Fraser fir (*Abies fraseri*). Because relatively few mountains achieve such stature, spruce-fir forests are restricted to just a few "sky islands" in the Great Smoky Mountains (NC and TN), the Balsam Mountains (NC), the Black Mountains (NC), and Mount Rogers (VA). Soils are shallow, highly organic, and acidic, providing limited nutrients and physical support. Winters are long and harsh, while the short summer growing seasons are relatively cool. Precipitation is abundant at high elevations; sources include rainfall in the summer months, fog drip year-round, and wind-driven snow and ice in winter.

In his classic description of the spruce-fir forests of the high mountains of North Carolina, B. W. Wells, in *Natural Gardens of North Carolina*, referred to these forests as "Christmas Tree Land" because of their resemblance to the extensive boreal forests to the north. Unfortunately, the dense, dark, aromatic, and stately stands of spruce and fir to which Wells referred have largely vanished, victims of an introduced insect pest, air pollution, and residual effects of extensive logging and severe fires in the late 19th and early 20th centuries. Consider yourself fortunate to find a patch of older spruce-fir forest to experience.

Although red spruce occurs in forests far to our north, Fraser fir is a species restricted to the southern Appalachian Mountains. Red spruce and Fraser fir have different elevation ranges in the southern Appalachians as well, with red spruce dominant at lower elevations. Fraser fir becomes a codominant at higher elevations, typically around 5,500 ft, and on the highest summits (6,000+ ft), it grows in nearly pure stands. Few other trees thrive in this harsh environment, but both yellow birch (*Betula alleghaniensis*) and mountain ash (*Sorbus americana*) can be found. Mountain maple (*Acer spicatum*) is an occasional understory tree, and fire cherry (*Prunus pensylvanica*) may occupy disturbed patches.

Understories are typically open (in stands with a closed canopy), and diversity is low, with wood fern (*Dryopteris campyloptera*), wood sorrel (*Oxalis montana*), and Canada mayflower (*Maianthemum canadense*) common, along with mosses (particularly feather moss, [*Hylocomium splendens*]). Frequent shrubs include witch hobble (*Viburnum lantanoides*) and southern mountain cranberry (*Vaccinium erythrocarpum*). Adelgid-damaged stands (see below) are often overgrown with blackberries (*Rubus* spp.) and invasive European grasses and forbs. Spruce-fir forest typically transitions gradually into northern hardwood forest or montane oak forest toward lower elevations, but it may also adjoin, typically across abrupt boundaries, shrub bald, grass bald, and rocky summit natural communities.

We know that the summits of the high southern mountains of 6,000+ ft have not always been forested. We have evidence (in the form of pollen from mountain bogs) that during the most recent Ice Age, which ended about 12,000 years ago, boreal forest covered the entire range of the southern Appalachians, except for the highest summits, which likely supported alpine tundra. As the climate warmed, boreal forest species found refuge in the mountains on the highest peaks, and tundra gradually disappeared. Southern Appalachian spruce-fir forests differ from their northern counterparts in terms of seasonal day-length patterns, milder winters and warmer summers, and species composition. Compared side by side, the spruce-fir forests in the northern and southern Appalachians share only about half of their species. On high mountains in the north, alpine tundra replaces spruce-fir forest above 4,000 ft. In the southern Appalachians, even the highest summits (which exceed 6,000 ft) are covered with spruce-fir forest. For further discussion of how the elevation of treeline (the upper limit of spruce-fir forest) varies with latitude, please refer to the sidebar in hike 20 (Mount Craig), "Treeline and Latitude."

Southern spruce-fir forests are severely threatened. Logging and the devastating effects of extremely hot slash fires associated with logging took their toll on these forests in the late 19th and early 20th centuries, leaving some sites permanently open after fires consumed the organic soils. Logging is no longer a serious threat, however, because most remaining spruce-fir forests are in protected areas, such as state and national parks. Since the 1950s, an insect introduced accidentally from Southeast Asia, the balsam woolly adelgid, has devastated most of the Fraser fir in the region and left many formerly magnificent stands as bleached skeletons. Young Fraser fir trees seem relatively resistant to the adelgid, but it remains to be seen whether or not the regenerating stands will successfully reproduce. Air pollution and encroaching exotic species also have severe effects on the remaining forests. Clearly, climate is the main factor driving species composition at high elevations, preventing all but the hardiest species from living in the harsh conditions. One has to wonder what the future holds for these beleaguered forests, with climate models projecting higher temperatures in the southern Appalachians. The ultimate question, still unanswered, remains: Will spruce-fir forest migrate ever upward, until it simply runs out of space and disappears altogether, just as tundra did following the Ice Age? For more information on climate change effects on spruce-fir forests, please refer to the sidebar in hike 23 (Roan High Bluff and Roan Gardens), "Climate Change on Sky Islands."

You can see spruce-fir forests on hike 16 (Carlos Campbell Overlook) and hike through these forests on hikes 20 (Mount Craig), 23 (Roan High Bluff and Roan Gardens), 24 (Grassy Ridge), 25 (Grandfather Mountain), and 29 (Mount Rogers).

Cove Forest

Hiking through an old-growth rich cove forest is one of the most memorable experiences for any visitor to the southern Appalachian Mountains. Diverse, cathedral-like stands of stately trees tower well over 100 ft above an open understory that features a stunning diversity of herbaceous plants. Cove forests occupy protected landforms, including broad, flat-bottomed valleys (coves) and steeper ravines, often extending well up the adjacent sheltered slopes, especially on north-facing aspects. They can be found across a wide range of elevations, from as low as 1,000 ft to 4,500 ft and occasionally higher.

Cove forest soils are generally deep, and soil moisture is abundant; soil fertility can vary, leading to differences in species composition. The more acidic, nutrient-poor soils that develop from felsic rocks, which include most granites and gneisses, support relatively species-poor forests ("acidic coves") with canopies featuring eastern hemlock (*Tsuga canadensis*), tulip-tree (*Liriodendron tulipifera*), sweet birch (*Betula lenta*), Fraser magnolia (*Magnolia fraseri*), silverbell (*Halesia tetraptera*), and white pine (*Pinus strobus*). Understories of acidic coves tend to be shrubby, with rosebay rhododendron (*Rhododendron maximum*) and mountain doghobble (*Leucothoe fontanesiana*) typically forming impenetrable thickets, especially along stream corridors.

Sites on more mafic rocks develop nutrient-rich, nearly neutral-pH soils that support the aforementioned species but also many others, with some sites featuring as many as 30 or more canopy species. These "rich coves" support the most species-rich forests in eastern North America. Noteworthy additional canopy species include basswood (*Tilia americana*), white ash (*Fraxinus americana*), yellow buckeye (*Aesculus flava*), cucumber magnolia (*Magnolia acuminata*), black cherry

(*Prunus serotina*), and several oak species (*Quercus* spp.). The herbaceous layer is remarkable for its diversity of spring wildflowers (before canopy leaf-out), when trilliums (*Trillium* spp.), spring beauties (*Claytonia* spp.), trout lily (*Erythronium americanum*), mayapple (*Podophyllum peltatum*), and many others are in full bloom. Blue cohosh (*Caulophyllum thalictroides*), common black cohosh (*Actaea racemosa*), wood-nettle (*Laportea canadensis*), bland sweet cicely (*Osmorhiza claytonii*), and yellow mandarin (*Prosartes lanuginosa*) are among the best herbaceous indicators of rich cove forests. Further discussion of herbaceous diversity in rich cove forests may be found in the sidebar in hike 6 (Station Cove Falls), "Wildflower Diversity in Temperate Forests."

Acidic and rich cove forests can coexist in coves and ravines that have complex geology and topography. At higher elevations on protected topography, cove forests intergrade with northern hardwood forests. Toward more open and exposed topographic positions, cove forests intergrade with oak forests or montane oak forests, depending on elevation.

Cove forests are limited in extent and critically important as reservoirs of biological diversity for both plants and animals. Coves of the southern Appalachians may have served as refuges for warm temperate flora during cooler climates. Many of the plant species of southern Appalachian cove forests have close relatives in the forests of eastern Asia, providing evidence of ancient connections between these areas. This interesting topic is explored in more depth in the sidebar for hike 14 (Joyce Kilmer Memorial Forest), "Eastern North American–Eastern Asian Connections."

Threats include logging, both past and present, which can convert species-rich, all-aged stands into less diverse even-aged stands. The chestnut blight in the early to mid-20th century eliminated American chestnut (*Castanea dentata*), an important tree in cove forests throughout the Appalachians. A current threat is the hemlock woolly adelgid, currently devastating eastern hemlock populations throughout their range. Hemlock woolly adelgid is particularly destructive in acidic cove forests, where eastern hemlock is often one of the dominant tree species. Further information about hemlock woolly adelgid may be found in the sidebar for hike 18 (Ramsey Cascades), "Hemlock Woolly Adelgid."

Many "flavors" of cove forest are featured in this book; see hikes

1 (Blood Mountain), 2 (Sosebee Cove), 4 (Tennessee Rock Trail), 6 (Station Cove Falls), 7 (Fork Mountain Trail), 9 (Pinnacle Mountain), 10 (Rainbow Falls), 14 (Joyce Kilmer Memorial Forest), 15 (Gregory Bald), 16 (Carlos Campbell Overlook), 17 (Alum Cave), 18 (Ramsey Cascades), and 25 (Grandfather Mountain).

Low-Elevation Pine Forest

Low-elevation pine forests of moderate stature (40 to 80 ft) are widespread in the southern Appalachians across a variety of topographic positions, including ridgetops, open slopes, and broad valleys, almost always below 2,700 ft elevation. Soils are typically acidic and nutrient-poor. Dominant tree species are shortleaf and Virginia pines (*Pinus echinata* and *P. virginiana*), although pitch pine (*P. rigida*) may also occur in low-elevation pine forests. Low-elevation pine forests tend to occupy larger patches and less extreme topographic positions than montane pine forest and woodland.

Other trees of low-elevation pine forests are mainly oaks, including chestnut, scarlet, and black (*Quercus montana*, *Q. coccinea*, and *Q. velutina*, respectively), plus broad-leaved species such as pignut hickory (*Carya glabra*), sourwood (*Oxydendrum arboreum*), sassafras (*Sassafras albidum*), black gum (*Nyssa sylvatica*), and red maple (*Acer rubrum*). The shrub layer is often dominated by ericaceous shrubs, with such species as low bush blueberry (*Vaccinium pallidum*), black (*Gaylussacia baccata*) and bear (*G. ursina*) huckleberries, mountain laurel (*Kalmia latifolia*), and rosebay rhododendron (*Rhododendron maximum*) common. Sprouts of the American chestnut (*Castanea dentata*) are abundant, hinting at its former importance in the canopy, and the shrubby chinquapin (*C. pumila*) may also be found. The herbaceous layer, when present, includes a variety of low-growing species like pipsissewa (*Chimaphila maculata*), trailing arbutus (*Epigaea repens*), wintergreen (*Gaultheria procumbens*), galax (*Galax urceolata*), and bracken fern (*Pteridium aquilinum*). Low-elevation pine forests typically grade into oak forests in more sheltered or otherwise mesic sites.

Low-elevation pine forests are typically even-aged, owing to their frequent establishment following fire. The most important current threat is the suppression of stand-renewing fires by human activities. Without fire, pine forests are eventually replaced by oak forests dominated by drought-tolerant oaks, hickories, sourwood, black gum, sassafras, and red maple. When these species take over a low-elevation pine forest, they create conditions that are less conducive to future fire, leading to long-term conversion of pine to broad-leaved forests. Please refer to the sidebar in hike 21 (Babel Tower and Wiseman's View), "Fire on the Mountain," for more information about the effects of fire on natural communities. Other serious threats to low-elevation pine forests in the southern Appalachians include timber harvest, development, prolonged drought, and outbreaks of southern pine beetle.

Two hikes in this book will take you through low-elevation pine forests, and you will find more of this natural community in the adjacent Piedmont. Hikes 5 (Martin Creek Falls) and 8 (Oconee Bells Nature Trail) feature some fine examples.

Montane Pine Forest and Woodland

At elevations above 2,300 ft, steep, rocky ridges on southerly to westerly aspects in the southern Appalachians often support patches of forest or woodland in which pitch (*Pinus rigida*) and/or Table Mountain (*P. pungens*) pines are dominant/codominant. Pitch pine is widespread in eastern North America, while Table Mountain pine is a southern Appalachian endemic.

Montane pine forests and woodlands have low to moderate stature (40 to 60 ft) and typically open canopies that admit plenty of sunlight to the understory. The distribution of montane pine forests and woodlands is dictated mainly by topography. The ridges and other exposed landforms where these natural communities occur have shallow, rocky soils typically developed from acidic bedrock. The acidic soils provide few available nutrients and have low water-holding capacity, owing to their shallow depth, sandy texture, and abundance of rocks. A thick "duff" of partially decomposed organic matter typically lies above the mineral soil and contains a high density of plant roots. Limited soil water-holding capacity and rapid drainage, combined with exposure to intense sun and high winds, result in a drought-prone environment. Lightning-ignited fires of varying intensity periodically sweep through this natural community.

Some stands may include Carolina hemlock (*Tsuga caroliniana*), which can create a denser, shadier canopy structure. Other common canopy species include drought-tolerant, broad-leaved hardwood species, such as scarlet (*Quercus coccinea*) and chestnut (*Q. montana*) oaks, as well as sourwood (*Oxydendrum arboreum*), black gum (*Nyssa sylvatica*), red maple (*Acer rubrum*), and sassafras (*Sassafras albidum*). The understory is often woody and dominated by ericaceous shrubs, including mountain laurel (*Kalmia latifolia*) and azaleas and rhododendrons (*Rhododendron* spp.), as well as blueberries (*Vaccinium* spp.) and huckleberries (*Gaylussacia* spp.). At higher elevations, Catawba rhododendron (*Rhododendron catawbiense*) may also be common in the understory. Because of the abundance of ericaceous shrubs in the understory of many montane pine forests and woodlands, these natural communities are sometimes referred to as pine-oak-heaths. Few low-growing species occur in the understory, but those that do are largely the same as those mentioned for the low-elevation pine forest. Montane pine forests and woodlands typically grade into montane oak forests in more sheltered or otherwise mesic sites and often surround granitic domes and rocky summits.

Fire is critical for regenerating montane pine forest and woodland in the southern Appalachians, and extensive fires can lead to even-aged stands following pine regeneration. Patchier, smaller disturbances, such as wind-throw, lightning strikes, and damage from ice storms, result in local regeneration and an uneven-aged forest. Pitch and Table Mountain pines have different adaptations to fire, discussed in detail in the sidebar in hike 22 (Table Rock), "A Tale of Two Pines."

Outbreaks of southern pine beetle without a stand-renewing fire kill pines and allow drought-tolerant species to take over, including oaks, red maple, and white pine (*P. strobus*). Once established, these mixed forests are less flammable than their pine-dominated counterparts, so a potentially permanent conversion from pine to hardwood dominance can occur. Restoration of such stands to montane pine forest and woodland is difficult and expensive. Recent droughts, combined with frequent outbreaks of southern pine beetle and active fire suppression, have led to regionwide decline in these attractive and ecologically important communities. Please refer to the sidebar in hike 21 (Babel Tower and Wiseman's View), "Fire on the Mountain," for more information about the effects of fire on natural communities.

As is the case with low-elevation pine forests, the most important current threat to montane pine forests and woodlands is the suppression of stand-renewing fires by human activities. Prolonged drought and outbreaks of southern pine beetle are likewise of great concern. Montane pine forests and woodlands are not threatened by timber harvest, but residential development is a potentially serious threat. Projected climate changes may benefit montane pine forests and woodlands, possibly expanding their range as temperatures rise and droughts become more pronounced in the region, which may increase the likelihood of fires that stimulate regeneration of montane pine forests and woodlands. However, increased monitoring and fire suppression by humans may offset this effect. Increased drought may also increase the likelihood of southern pine beetle outbreaks, another factor acting in opposition to the expansion of montane pine forests and woodlands.

You can see the die-off of montane pine forests and woodlands from hike 16 (Carlos Campbell Overlook). Active restoration efforts for this forest type can be observed on hike 15 (Gregory Bald), while Linville Gorge has two excellent examples on hikes 21 (Babel Tower and Wiseman's View) and 22 (Table Rock), which burned in a 2013 fire.

Northern Hardwood Forest

Hikers familiar with the upper midwestern and northeastern United States, and the adjacent Canadian provinces, will have no doubt walked through deciduous forests dominated by American beech (*Fagus grandifolia*), yellow birch (*Betula alleghaniensis*), and sugar maple (*Acer saccharum*). These so-called beech-birch-maple forests are common in settings of intermediate moisture at low to mid-elevations, occurring as far west as Minnesota and as far south as Pennsylvania. Similar forests also occur at higher elevations (4,000 to nearly 6,000 ft) in the southern Appalachians and are thus referred to as "northern hardwoods." The best examples are magnificent stands of tall (80 to 100 ft) trees, all-aged, including some of substantial age and girth. These forests are best-developed on north- and east-facing slopes on a wide variety of rock types. The climate of these forests is cool and moist but generally not subject to the harsh winter conditions of the highest elevations. Soils are rocky, well-drained, and moist, with varying fertility. Extreme rockiness characterizes high-elevation boulderfields.

Although similar in some respects to their northern counterparts, our southern "northern" hardwood forests are unique, owing to a milder climate and longer growing season. In addition, the sites occupied by these forests were never buried beneath glaciers. Our northern hardwood forests usually have American beech, yellow birch, and sugar maple found in their namesake forests, but they may also include yellow buckeye (*Aesculus flava*) along with black cherry (*Prunus serotina*), basswood (*Tilia americana*), red oak (*Quercus rubra*), and red spruce (*Picea rubens*). American chestnut (*Castanea dentata*) was important in pre-blight forests. Logging and severe fire may stimulate recruitment of rapidly growing but short-lived fire cherry (*Prunus pensylvanica*) and black locust (*Robinia pseudoacacia*). An understory of smaller trees and shrubs is typically present, including striped (*Acer pensylvanicum*) and mountain (*A. spicatum*) maples, smooth serviceberry (*Amelanchier laevis*), flame azalea (*Rhododendron calendulaceum*), and wild hydrangea (*Hydrangea arborescens*). A herbaceous layer (sometimes quite species-rich) is usually present, including many forbs also found in rich cove forests. Some stands are carpeted with ferns; others have a dense sward of the fine-leaved Pennsylvania sedge (*Carex pensylvanica*). Like its northern cousin, this natural community is ablaze with brilliant leaf colors during the fall.

The northern hardwood forest adjoins several other natural communities, grading into spruce-fir forest, montane oak forest, or cove forest, depending on the direction of environmental change. The transition to cove forest is often gradual and extensive, making the boundary between these natural communities especially indistinct. Other natural communities, such as grass bald, shrub bald, seepage wetland, granitic dome, and rocky summit, may be embedded within northern hardwood forest. High-elevation boulderfields may be dominated by yellow birch, and high-elevation "gaps" (low places along ridgelines) may be dominated by American beech; both are specialized examples of northern hardwood forest. Some ecologists would argue that northern hardwood forests should not be labeled a distinct community at all, being at the intersection of many other communities. To learn more about this debate, see the sidebar in hike 26 (Elk Knob), "Northern Hardwoods—Natural Community or Crossroads?"

Most northern hardwood forests appear to be relatively stable, with smaller, local disturbances creating forest gaps and stimulating regeneration. Larger-scale disturbances, such as fire or extreme weather events, are relatively rare but do occasionally affect this natural community. Threats include logging and development of unprotected natural areas. Effects of climate change are also serious threats, with drought and increased likelihood of devastating fire as the two most serious consequences of a warming climate for northern hardwood forests.

Northern hardwood forests are widespread, with half the hikes in this book including one or more subtypes. Hike 26 (Elk Knob) has all three subtypes of northern hardwood forest. Hikes 3 (Brasstown Bald) and 4 (Tennessee Rock Trail) have both the typical mixed forest as well as boulderfield forests. Hikes 12 (Whiteside Mountain), 13 (Hooper Bald), 15 (Gregory Bald), 16 (Carlos Campbell Overlook), 17 (Alum Cave), 18 (Ramsey Cascades), 19 (Craggy Pinnacle), 25 (Grandfather Mountain), 28 (Whitetop Mountain), 29 (Mount Rogers), and 30 (Twin Pinnacles Trail) all contain the more typical mixture of northern hardwood species.

Oak Forest

The most extensive natural community of the southern Appalachians is the oak forest. These are medium- to tall-stature (60 to 80 ft), closed-canopy forests typically dominated by a variety of oaks, including chestnut (*Quercus montana*), scarlet (*Q. coccinea*), white (*Q. alba*), red (*Q. rubra*), and black (*Q. velutina*) oaks. Identifying the dominant oaks is important, so please refer to the sidebar in hike 7 (Fork Mountain Trail), "Southern Appalachian Oaks," for more information about oaks of the region.

Other tree species that are common in oak forests include hickories (*Carya* spp.), black gum (*Nyssa sylvatica*), red maple (*Acer rubrum*), white pine (*Pinus strobus*), sourwood (*Oxydendrum arboreum*), and white ash (*Fraxinus americana*). American chestnut (*Castanea dentata*) was a dominant or codominant in oak forests until its elimination by chestnut blight during the early 1900s. Oak forests of the southern Appalachians are different from adjacent Piedmont forests owing to the presence of montane species, such as Fraser magnolia (*Magnolia fraseri*), bear huckleberry (*Gaylussacia ursina*), and flame azalea (*Rhododendron calendulaceum*).

These forests occur on a wide variety of topographic positions and aspects at low to mid-elevations, typically below about 3,000 ft. Underlying rocks and resultant soils are also highly varied, so the oak forest concept is rather broad. Although old-growth oak forests are typically all-aged, hikers are more likely to encounter even-aged forests recovering from past logging. In more protected ravines and coves, oak forest grades into cove forest, and on more exposed south-facing slopes and steep ridges, oak forest grades into either low-elevation pine forest or montane pine forest and woodland. Above 3,000 ft, oak forest blends into montane oak forest. Because oak forests occur on a wide range of rock and soil types, there is considerable variation in understory species, which typically include either deciduous (blueberry [*Vaccinium* spp.], huckleberry [*Gaylussacia* spp.], azalea [*Rhododendron* spp.]) or evergreen (mountain laurel [*Kalmia latifolia*] or rosebay rhododendron [*Rhododendron maximum*]) ericaceous shrubs. Herbaceous species are sparse to moderately abundant and highly varied from site to site, including many forbs, grasses, and ferns.

The loss of codominant American chestnut to chestnut blight during the early 1900s altered oak forests in fundamental ways. Widespread logging (including salvage logging of American chestnut) has destroyed or altered many oak forests. More information about the chestnut blight and its consequences may be found in the sidebar for hike 27 (Mount Jefferson), "Where Has the Chestnut Gone?"

Successional stands established following logging or abandonment of cultivation/pasture are often dominated by widely dispersed and fast-growing species, such as white, shortleaf (*Pinus echinata*), and Virginia (*P. virginiana*) pines; tulip-tree (*Liriodendron tulipifera*); and black locust (*Robinia pseudoacacia*).

One of the greatest threats to the integrity of oak forests is invasion by exotic species, such as chestnut blight. Exotic invasive insects, such as gypsy moth and emerald ash borer, pose serious threats to oak forests, as do exotic invasive plants, such as princess tree (*Paulownia tomentosa*) and tree of heaven (*Ailanthus altissima*). Sudden oak death, a fungal disease of uncertain origin, has established itself in parts of the western U.S., but it has not yet spread to eastern forests; its effects on southern Appalachian oak forests would be devastating. Other threats include continued logging and land conversion to nonforest uses. Fire suppression is also an ongoing concern, as it results in poor oak regeneration and eventual replacement of oak forests by forests of mesophytic, fire-intolerant species, such as red maple, tulip-tree, black gum, and white pine.

Oak forests are the most common forests in the southern Appalachians, and they cover the greatest land area. Hikes through oak forests in this book include 5 (Martin Creek Falls), 7 (Fork Mountain Trail), 8 (Oconee Bells Nature Trail), 9 (Pinnacle Mountain), 15 (Gregory Bald), and 21 (Babel Tower and Wiseman's View).

Grass Bald

Grass balds are open, grassy meadows occurring sporadically at higher elevations (typically above 4,500 ft) throughout the southern Appalachians. They occupy a variety of rock and soil types and topographic positions, but soils are generally rocky and shallow. Grass balds are typically fringed by shrub balds, montane oak forests, northern hardwood forests, or spruce-fir forests. Because even the highest elevations in the southern Appalachians (6,000+ ft) are well below the elevation (treeline) where climate would dictate more open, alpine vegetation, the cause of open, grassy areas within the forested landscape has been a subject of speculation. What is clear is that most grass balds are subject to invasion by shrubs and trees and will eventually revert to forest unless artificially maintained by mowing, clipping of woody plants, or grazing. Grass balds are featured in two sidebars in this book, in hike 13 (Hooper Bald), "Grass Bald Origins and Future," and in hike 29 (Mount Rogers), "When a Grass Bald Isn't."

Grass balds are most often found along broad ridges and on gentle, rounded summits at high elevations. However, they occupy only a small fraction of the available habitat; this fact suggests that their origin and maintenance require more than a particular set of environmental conditions. Current thinking about grass balds leans heavily toward the important role of large grazing mammals in their maintenance, possibly native animals (including some that are now extinct) that first began grazing alpine grasslands above treeline during the glacial maximum. In postsettlement time, there is good evidence (including historical photographs like the one of Craggy Pinnacle, hike 19) that domestic grazing animals were pastured on grass balds during the summer months and that their grazing helped keep the balds open. However, other hypotheses have been proposed, and the origin and presettlement maintenance of these attractive and floristically diverse natural communities remains a mystery.

Grass balds are typically dominated by mountain oat grass (*Danthonia compressa*) and other grasses, Pennsylvania sedge (*Carex pensylvanica*) and other sedges, and many forbs, including Michaux's saxifrage (*Hydatica petiolaris*), goldenrods (*Solidago* spp.), three-toothed cinquefoil (*Sibbaldia tridentata*), wild strawberry (*Fragaria virginiana*), and many others. Brambles (*Rubus* spp.) are common, along with scattered ericaceous shrubs (particularly Catawba rhododendron [*Rhododendron catawbiense*], flame azalea [*R. calendulaceum*], and blueberries [*Vaccinium* spp.]). Small trees are common around the edges of the grass balds, and these include familiar species such smooth serviceberry (*Amelanchier laevis*), fire cherry (*Prunus pensylvanica*), white pine (*Pinus strobus*), and red spruce (*Picea rubens*). Grass balds may also be overrun with numerous exotic pasture grasses, most of these imported from Europe. There are relatively few rare species associated with grass balds, although showy Gray's lily (*Lilium grayii*) is a well-known example.

The most important current threat to grass balds is invasion by woody plants in the absence of active management. Most grass balds will disappear without human intervention to limit woody invasion. Invasion by exotic species, particularly grasses, is an ongoing threat. Wild boar, an invasive exotic animal, can also be problematic, because they uproot large areas in search of tasty belowground plant parts and small animals. Residential development is a threat in some areas. Because of their historical interest and scenic beauty, several well-known grass balds (such as Gregory and Hooper Balds) are being actively managed (with mowing and grazing) by their landowners. In our book, we feature managed grass balds in hikes 13 (Hooper Bald), 15 (Gregory Bald), 24 (Grassy Ridge), and 28 (Whitetop Mountain).

Shrub Bald

If you've driven through the Tennessee side of Great Smoky Mountains National Park on US 441, you've probably noticed odd-looking patches of treeless vegetation on steep, south-facing slopes and ridges above you. These high-elevation treeless patches (most of them above 4,500 ft) appear meadowlike and smooth-textured from a distance; thus these natural communities are locally referred to as "balds" or "slicks." On closer examination the hiker will discover that these are actually dense thickets of shrubs in the Ericaceae family, leading to the origin of another name for these places: "heath balds." You can learn more about several of the key species of shrub balds in the sidebar for hike 1 (Blood Mountain), "Rhododendrons and Mountain Laurel."

Like grass balds, shrub balds occupy places in the landscape that would be expected to support high-elevation forests, such as northern hardwood forest and spruce-fir forest. Shrub balds are associated with steep, south-facing slopes on unstable acidic rocks with shallow, organic soils over bedrock. Extreme weather, the poor rooting environment, the strongly competitive nature of the shrubby vegetation, and possibly fire all contribute to the maintenance of shrubby vegetation to the exclusion (or near-exclusion) of trees. Unlike grass balds, shrub balds occupy a more predictable place in the environment, and they are not dependent on human intervention for their maintenance.

Catawba rhododendron (*Rhododendron catawbiense*) is a typical dominant in shrub balds, but other evergreen rhododendrons (rosebay rhododendron [*R. maximum*] and gorge rhododendron [*R. minus*]) are also important. Other shrubs and small trees found in shrub balds include mountain laurel (*Kalmia latifolia*), sand myrtle (*K. buxifolia*), highbush blueberries (*Vaccinium* spp.), black chokeberry (*Aronia melanocarpa*), mountain sweet pepperbush (*Clethra acuminata*), fire cherry (*Prunus pensylvanica*), mountain ash (*Sorbus americana*), and red spruce (*Picea rubens*). Herbaceous species are essentially absent in shrub balds.

There are few known threats to shrub balds, and changing climate could contribute to their stabilization and possible expansion, if increased drought and more frequent fire limit the successful invasion of forest trees. You can see a shrub bald from hike 16 (Carlos Campbell Overlook) and hike through examples on hikes 17 (Alum Cave), 19 (Craggy Pinnacle), 23 (Roan High Bluff and Roan Gardens), 24 (Grassy Ridge), and 27 (Mount Jefferson).

Seepage Wetland

Steeply sloping rock surfaces with perennial water seepage support natural communities that occur as small, scattered patches or as longer, vertical features. These seepage wetlands are found over a wide range of elevations but are most common at higher elevations where cooler temperatures and higher precipitation favor their development. The vegetation is predominantly herbaceous, and it varies from continuous plant cover over saturated soils to smaller patches of vegetation separated by areas of bare rock, boulders, gravel, or exposed mineral soil. The stature of the vegetation is generally low (below 4 ft), but seepage wetlands are often bordered by areas of shrubby vegetation and may have substantial tree cover from adjacent forested communities.

These communities are too wet for trees and woody species, and ice formation and slippage at higher elevations in the winter can remove all established vegetation. Vegetation is highly variable and potentially diverse, including (in the examples covered in this book) showy wildflowers such as touch-me-nots (*Impatiens* spp.), meadow-rues (*Thalictrum* spp.), Appalachian bluet (*Houstonia serpyllifolia*), Turk's cap lily (*Lilium superbum*), wood-nettle (*Laportea canadensis*), St. John's-worts (*Hypericum* spp.), Michaux's (*Hydatica petiolaris*) and other saxifrages, deerhair bulrush (*Trichophorum cespitosum*), grasses, sedges, ferns, and *Sphagnum* mosses. Many rare species occur in this natural community; the lengthy list includes several rare sedges, grass-of-parnassus (*Parnassia* spp.), and both fringed and fringeless orchids (*Platanthera* spp.). Numerous rare ferns, mosses, and liverworts are also found in seepage wetlands.

Seepage wetlands can be embedded within any of the mid- to high-elevation natural communities in the southern Appalachians, including cove forest, northern hardwood forest, spruce-fir forest, oak forest, montane oak forest and woodland, grass bald, shrub bald, granitic dome, and rocky summit.

Although they are distinctive natural communities, seepage wetlands are closely related by their environment, flora, and vegetation to two communities we have chosen not to address in this book, spray cliffs (natural communities of perennially moist rock faces associated with waterfalls) and bogs and fens (natural communities of perennially saturated soils on relatively level sites). The seepage wetland and its cousins are fragile communities that are extremely sensitive to damage by human foot traffic (so please observe them from a distance), and they can be damaged by activities associated with logging and development, including changes in hydrology.

Climate change will likely impact seepage wetlands, including increased scouring by flash floods associated with extreme rainfall events, prolonged droughts and higher temperatures, and invasions of exotic species that already threaten lower-elevation seepage wetlands. An indirect effect is the devastation of eastern hemlock (*Tsuga canadensis*) by an invasive exotic insect, hemlock woolly adelgid. Hemlocks provide dense shade for adjacent mountain wetlands, and their decline will subject these wetlands to potentially harmful effects of greater sunlight and higher temperatures.

Although not specifically covered in this book, Wolf Mountain Overlook (5,500 ft), at milepost 424.8 on the Blue Ridge Parkway, is one of our favorite examples of a high-elevation seepage wetland, located directly across the highway from the overlook. This extensive and accessible rock face is also a fine example of the successful blending of art and science in landscape architecture, having been intentionally created during construction of the Blue Ridge Parkway to resemble a natural seepage wetland. Seepage wetlands are also featured in hike 12 (Whiteside Mountain).

Granitic Dome

A granitic dome, as the name suggests, is a rounded outcropping of granite or related rock, typically standing well above the surrounding terrain above 3,000 ft. The domes, which are largely bare rock, originated in ancient times when magma welled up beneath older rocks but failed to break through to the earth's surface. As these great masses of molten material cooled, they became blobs of crystalline rock called plutons. The forces that formed these plutons cause the rocks to fracture along lines parallel to their surfaces, so weathering removes thin sheets or layers of rock, a process called spalling or exfoliation. Exfoliation is much like the process of peeling away a thin layer from an onion, which leaves another smooth surface without vertical fissures.

Granitic domes have smooth, impenetrable rock surfaces, which pose special challenges to plants. Soil is found only in relatively level areas and small depressions, where the initial growth of hardy lichens and mosses can trap mineral and organic particles, beginning the process of soil formation. These soils are shallow and drought-prone, offering little in the way of physical support, water, and nutrients, the very things that most plants require from soils. Disturbances such as severe storms can sweep away mats of vegetation and thin soils, leaving the surface bare once again. Even in areas where soils remain, the physical environment is unforgiving, with extreme exposure to sun, wind, severe storms, and rapidly changing temperatures.

The visitor to a granitic dome is rewarded with exceptional views in fair weather, and the domes themselves are spectacular to view from afar. The natural community is comprised of hardy plants that can thrive in such extreme environments. There are some interesting lichens, such as the Carolina rocktripe (*Umbilicaria caroliniana*) and the common toadskin (*Lasallia papulosa*). Mosses are also commonly found; some of the mosslike plants aren't mosses at all, however, but lower vascular plants called spikemosses (genus *Selaginella*). One common spikemoss, *Selaginella tortipila*, is so named because its tiny leaves terminate in long, twisting hairs. Flowering plants that grow on the granitic domes include numerous grasses and sedges, along with showier species, such as mountain dwarf-dandelion (*Krigia montana*), granite dome St. John's-wort (*Hypericum buckleyi*), southern bush honeysuckle (*Diervilla sessilifolia*), mountain-mints (*Pycnanthemum* spp.), and the granite dome bluet (*Houstonia longifolia* var. *glabra*). For further discussion of species of extreme habitats, please refer to the sidebar for hike 25 (Grandfather Mountain), "Extreme Botany."

Most granitic domes are fringed with shrubby vegetation dominated by ericaceous species typical of shrub balds. Common species include Catawba rhododendron (*Rhododendron catawbiense*), rosebay rhododendron (*R. maximum*), mountain laurel (*Kalmia latifolia*), and sand myrtle (*K. buxifolia*). Beyond the shrub zone, deeper soils may support pines, notably pitch pine (*Pinus rigida*) and drought-tolerant oaks. Surrounding communities include cove forest, oak forest, montane oak forest, and montane pine forest and woodland.

The principal threat to granitic domes is damage to the fragile plant communities by human traffic, notably hikers, picnickers, and rock climbers. Take care not to trample the mats of herbaceous species and lichens. Climate change is also considered a serious threat to the granitic dome natural community. Many granitic dome plant species are adapted to cool temperatures; increasing temperatures and more frequent droughts may eliminate them. In a changing climate, most granitic dome plant populations also have little opportunity for migration, either to higher elevations or to cooler environments to the north, so local extinctions are likely.

Granitic domes, such as hikes 11 (Satulah Mountain) and 12 (Whiteside Mountain), are especially abundant in the southeastern Blue Ridge Escarpment along the border between North and South Carolina but can also be seen on hike 1 (Blood Mountain) in Georgia.

Rocky Summit

Rocky summits are treeless rock outcrops that occur over a wide range of elevations and rock types throughout the southern Appalachians. They differ from granitic domes in having more fractured and irregular surfaces that provide opportunities for formation of stable pockets of soil. Like granitic domes, rocky summits represent harsh environments for plants, featuring drought, extreme weather, rapidly changing temperatures, and exposure to severe storms, high winds, and intense solar radiation. Not surprisingly, much of the rocky summit habitat consists of exposed rock with little vegetation other than hardy lichens and mosses. Rocky summits share some plant species with granitic domes (including mountain dwarf-dandelion [*Krigia montana*]), but they also have characteristic species, including Michaux's saxifrage (*Hydatica petiolaris*), wretched sedge (*Carex misera*), and a southern Appalachian endemic, rock-alumroot (*Heuchera villosa*). Joining these plants and adding to the colorful flora of the rocky summits are species that are widespread in other open or disturbed habitats in the southern Appalachians, including mountain oat grass (*Danthonia compressa*), common little bluestem (*Schizachyrium scoparium*), hairy five-fingers (*Potentilla canadensis*), and common bluet (*Houstonia caerulea*).

The highest rocky summits are of special interest because they harbor rare plants that have close ties to species of the arctic-alpine regions of northern New England and Canada. Some species, like cliff avens (*Geum radiatum*), are endemic to the southern Appalachians but have close relatives in other arctic-alpine habitats in the Northern Hemisphere. Other rare plants are northern disjuncts, including Greenland sandwort (*Minuartia groenlandica*), three-toothed cinquefoil (*Sibbaldia tridentata*), and deerhair bulrush (*Trichophorum cespitosum*). These plants occur in open habitats at high elevations in the southern Appalachians and then again only in alpine or tundra vegetation far to the north.

Scientists have pondered the origin of such disjunct distributions. One plausible explanation is that these are glacial relicts, plant species that were more widespread in the southern mountains during the last Ice Age (which ended about 12,000 years ago). Although the high peaks of the southern mountains were never glaciated, they supported large open areas of tundralike vegetation. As the postglacial climate warmed, the arctic-alpine species moved north, following the retreating glaciers, but some clung to the remaining cool and open habitats that remained in the southern Appalachians, such as rocky summits. These glacial relicts are rare in the South because of the very restricted nature of their habitats. They are important because they maintain a genetic record of flora that repopulated northern regions following glacial retreat. The sidebar for hike 11 (Satulah Mountain), "Ice Age Relicts," provides further discussion about glacial relicts.

Rocky summits are fragile and protected habitats, and land managers, including the National Park Service and various state parks agencies, have worked to protect these areas from damage caused by human traffic (hikers, picnickers, rock climbers). Please use caution when traversing these steep and rugged areas, and always stay within areas designated for human foot traffic. Aside from human disturbance, the plants of rocky summits are threatened by air pollution, encroachment of surrounding forest vegetation, and the direct effects of warming climate. You can visit rocky summits on hikes 19 (Craggy Pinnacle), 23 (Roan High Bluff and Roan Gardens), and 25 (Grandfather Mountain).

The Hikes

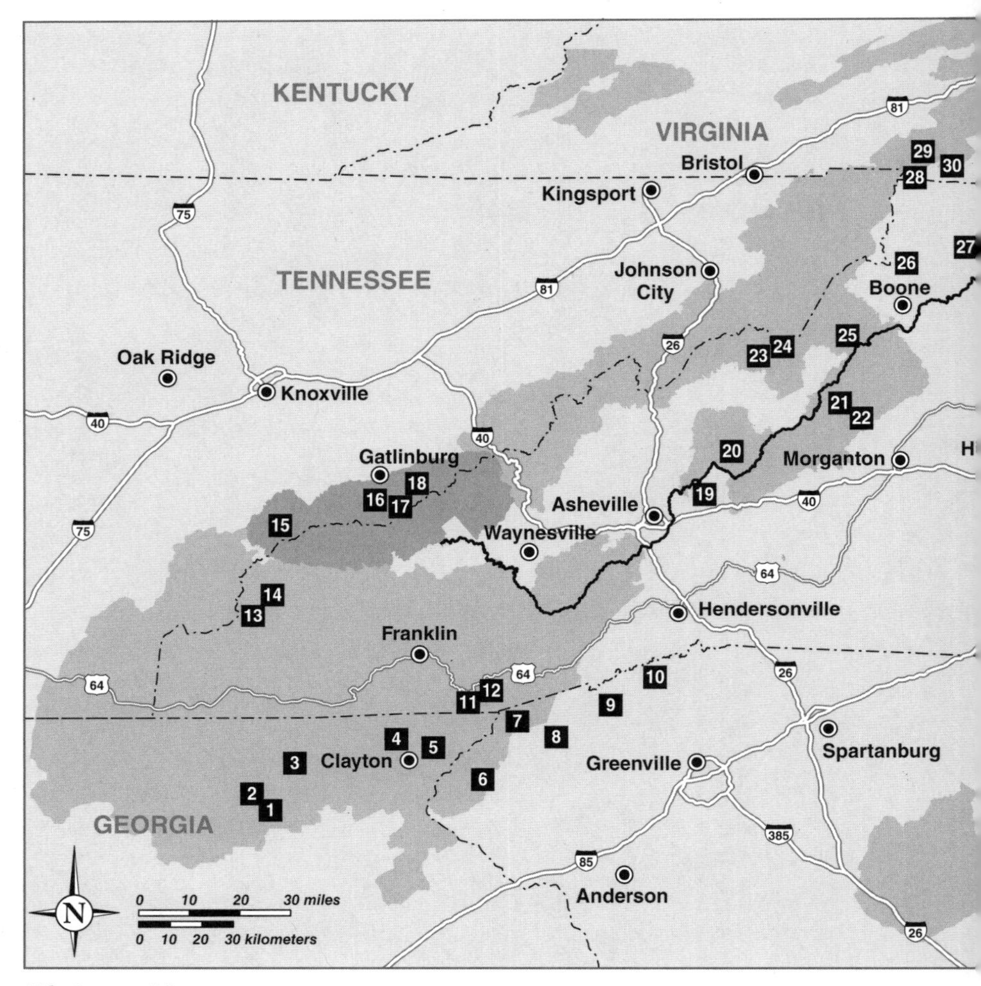

Hike Locator Map

1. Blood Mountain
2. Sosebee Cove
3. Brasstown Bald
4. Tennessee Rock Trail
5. Martin Creek Falls
6. Station Cove Falls
7. Fork Mountain Trail
8. Oconee Bells Nature Trail
9. Pinnacle Mountain
10. Rainbow Falls
11. Satulah Mountain
12. Whiteside Mountain
13. Hooper Bald
14. Joyce Kilmer Memorial Forest
15. Gregory Bald
16. Carlos Campbell Overlook
17. Alum Cave
18. Ramsey Cascades
19. Craggy Pinnacle
20. Mount Craig

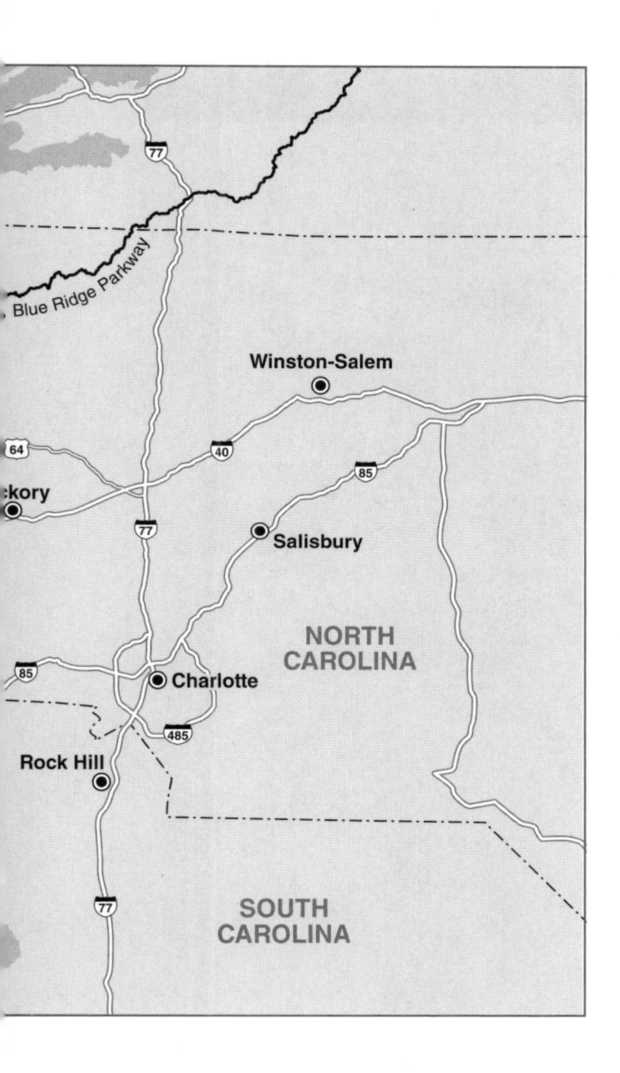

21. Babel Tower and Wiseman's View
22. Table Rock
23. Roan High Bluff and Roan Gardens
24. Grassy Ridge
25. Grandfather Mountain
26. Elk Knob
27. Mount Jefferson
28. Whitetop Mountain
29. Mount Rogers
30. Twin Pinnacles Trail

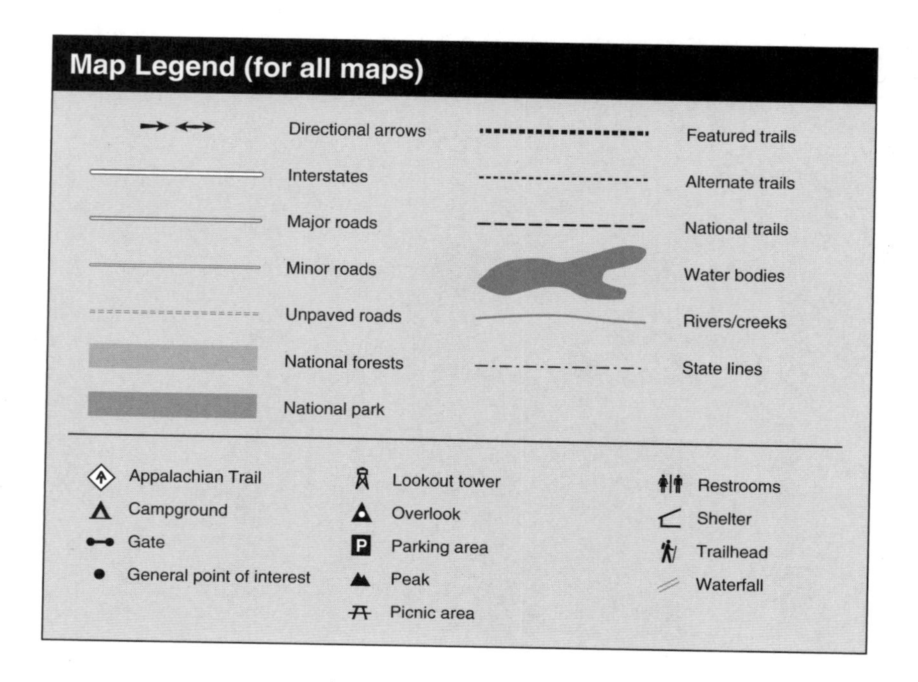
Map Legend (for all maps)
Directional arrows
Interstates
Major roads
Minor roads
Unpaved roads
National forests
National park
Featured trails
Alternate trails
National trails
Water bodies
Rivers/creeks
State lines
Appalachian Trail
Campground
Gate
General point of interest
Lookout tower
Overlook
Parking area
Peak
Picnic area
Restrooms
Shelter
Trailhead
Waterfall

1. BLOOD MOUNTAIN

Why Are We Here?

Blood Mountain, Slaughter Gap . . . the tranquil views you see at the top of Blood Mountain hardly suggest the violent skirmishes between the Creek and the Cherokee in the late 17th century. However, you may feel as if you've shed more than sweat by the time you reach the summit! If you take your time on the ascent, though, you'll see some important yet subtle changes in the forest, including variations in the composition, height, and cover. You'll learn to train your eyes to detect compositional modifications in vegetation that are linked to changes in the physical environment.

Blood Mountain is a popular and rewarding day hike. With its proximity to Atlanta and status as the highest point along the Appalachian Trail in Georgia, it is often crowded on nice weekends despite the elevation gain of approximately 1,300 ft to reach the Civilian Conservation Corps (CCC) shelter at the summit. Lace up your boots and carry some water!

The Hike

Blood Mountain is located in the Chattahoochee National Forest and within the Blood Mountain Wilderness. The hike begins at the clearly marked trailhead with a series of well-spaced blue blazes that define the Byron Herbert Reece Trail, which climbs 0.7 mi to intersect the Appalachian Trail at Flatrock Gap. This trail follows and crosses Shanty Branch several times during the first half mile before leaving the creek and climbing through a northeast-facing cove forest.

The composition of the natural community changes depending on its proximity to Shanty Branch, and you'll notice this as you cross its drainage several times. Along the creek you'll see an abundance of dying eastern hemlocks (*Tsuga canadensis*), which, like hemlocks on the entire East Coast, are infested with the invasive exotic insect hemlock woolly adelgid (*Adelges tsugae*). To learn more about the recent and catastrophic loss of eastern hemlock from our forests, please see our "Hemlock Woolly Adelgid" sidebar in hike 18 (Ramsey Cascades).

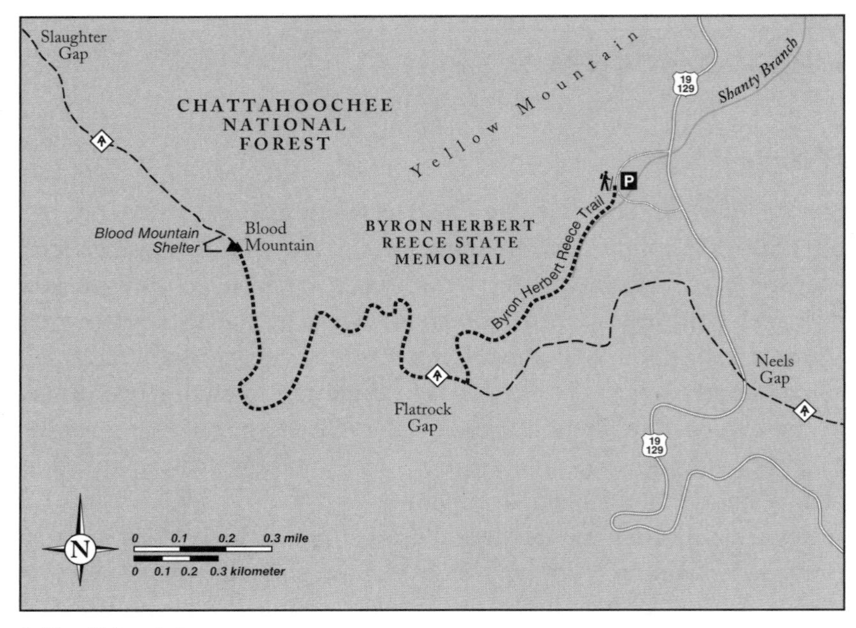

1. Blood Mountain

As is the case in many cove forests, the land forms a bowl shape, and its northeast aspect means that the trail stays relatively cool and moist during the hot Georgia summers. The forest understory is comprised of rosebay rhododendron (*Rhododendron maximum*) and mountain laurel (*Kalmia latifolia*). If these two evergreen shrubs are unfamiliar to you, please read the sidebar. Although there's not much to be seen under dense patches of these shrubs, you'll find a few open slopes with spring wildflowers as you climb the switchbacks up and away from the creek.

As you begin your climb away from the creek, you might want to begin a list of tree species, a useful practice on any hike. To eastern hemlock you can add several species of oaks (chestnut [*Quercus montana*], northern red [*Q. rubra*], and white [*Q. alba*]) plus red maple (*Acer rubrum*), sweet birch (*Betula lenta*), and tulip-tree (*Liriodendron tulipifera*). The understory here has some wildflowers in early spring through summer and fall, but if you count the number of species (of either trees or wildflowers), you'll find that the diversity is fairly low. Ecologists would attribute this to the acidic soil that underlies the Blood Mountain Wilderness. Why does an acidic cove have relatively few

1. BLOOD MOUNTAIN

Highlights: Challenging climb to a scenic view and historic stone CCC shelter; subtle changes in natural communities as you climb

Natural Communities: Cove forest (acidic subtype), montane oak forest, granitic dome

Elevation: 3,040 ft (trailhead); 4,461 ft (Blood Mountain summit)

Distance: 4.4 mi round-trip

Difficulty: The climb to the summit is moderately steep and rocky in places.

Directions: From Blairsville, GA, take US 19/129 south for 12 mi, then turn right into the Byron Herbert Reece Memorial and parking area (no facilities).

From Turners Corner, GA, where US 19 joins US 129, continue north on US 19/129 for approximately 8.3 mi. Just beyond the Walasi-Yi Center, take the second left into the Byron Herbert Reece Memorial and parking area (no facilities).

The Byron Herbert Reece parking area is limited to just 38 vehicles; if you arrive to find the lot full, continue south on SR 19/129 for a half mile, where you can access the Appalachian Trail near Neels Gap and hike the 2.4 mi to the Blood Mountain summit.

GPS: Lat. 34° 44′ 53.2″ N; Long. 83° 55′ 35.3″ W (trailhead)

Information: Blue Ridge Ranger District, 2042 Highway 515 West, Blairsville, GA 30512, (706) 745-6928

wildflowers? Acidic soils mean that there are fewer nutrients, such as calcium and magnesium, available for plants and a lower pH, which some plants can't tolerate.

If you've hiked through a variety of cove forests in the southern Appalachians, you may have noticed that there are several clues that indicate the acidity of the soils. In addition to the low species diversity, you'll typically find an abundance of species in the blueberry family, such as rhododendrons, mountain laurel, blueberries (*Vaccinium* spp.), and/or huckleberries (*Gaylussacia* spp.), and a canopy that includes eastern hemlock. Interestingly, acidic coves tend to self-perpetuate; the deep shade cast by the evergreen hemlocks and rhododendron plus the accumulation of acidic leaf litter on the ground prevent the establishment of other species. You may wonder what the forest will look like as the eastern hemlocks vacate the canopy and allow light to penetrate to the understory below. You share this question with many ecologists concerned for the future of these forests.

At 0.7 mi you will arrive at Flatrock Gap, where the Appalachian Trail crosses. Your options are to continue to follow the blue blazes

Rhododendrons and Mountain Laurel

Even the most casual hiker in southern Appalachian forests soon becomes intimately familiar with the region's "dynamic duo" of tall, evergreen, ericaceous shrubs: rosebay rhododendron (*Rhododendron maximum*) and mountain laurel (*Kalmia latifolia*). Covering several million acres in the southern Appalachians, these two species are beloved by wildflower enthusiasts, owing to their spectacular floral displays (mountain laurel in May and rosebay rhododendron in June). These same species are despised by foresters for their interference with regeneration of valuable timber species and cursed by cross-country orienteers who must bash their way through dense, impenetrable thickets of one or the other of these two species. The local term "rhododendron hell" for a thicket of rosebay rhododendron aptly sums up the latter situation.

It's well worth the effort to learn a little about the ecological distributions of these two important species. Like most of the ericaceous clan (species in the blueberry or heath family, Ericaceae), both species show an affinity for acidic soils. This is something of a chicken-and-egg scenario, because these species are also well-known for acidifying the soil through their contributions of abundant, base-deficient litter. Rosebay rhododendron favors more protected sites, such as acidic coves, where it forms dense stands along streams and up the adjacent protected slopes. Mountain laurel is generally found higher on the upper slopes and ridges, where it appears to favor shallower, rocky, and drought-prone sites. However, mountain laurel can also thrive on the same sites favored by rosebay rhododendron, which may outcompete it in such settings.

On the Blood Mountain hike, you'll observe that rhododendron is preva-

straight across to what is now called the Freeman Trail; turn left on the northbound, white-blazed Appalachian Trail to descend to Neels Gap; or turn right to follow the rectangular white blazes on the southbound Appalachian Trail to the summit of Blood Mountain. The climb to the summit is well worth your effort, so we recommend that you turn right to continue to the Blood Mountain shelter.

The first section of the trail climbs, then curves right to follow the contour before switchbacking steeply to climb above 4,000 ft. The

lent at lower elevations along the creek, nearly disappears along the mid-part of the hike, but reappears in open areas of rock outcrop below the summit. While this might appear to be a disjunct distribution, what you really see is two different species with different environmental requirements. Examine the rhododendron near the summit. You'll notice that instead of the gradually tapering triangular base of rosebay rhododendron, this shrub's evergreen leaves have a rounded base and an oval shape. This is Catawba rhododendron (*Rhododendron catawbiense*), which is found only in the higher elevations in the southern Appalachians and a few odd places in the Piedmont with cool microclimates.

Evergreen rhododendrons and mountain laurel have ecosystem roles that transcend their undisputable positions as the signature shrubs of southern Appalachian forests. They sequester (or store) vast amounts of carbon and mineral elements in their leafy and woody tissues. They regenerate from belowground rootstocks following the ravages of fire, windstorm, or chainsaws, quickly restoring plant cover and nutrient cycling to damaged sites. Indeed, these shrubs benefit from just about any process that opens the canopy above. Many extensive stands of rosebay rhododendron and mountain laurel probably regenerated following the last extensive fires (before widespread fire suppression) in the early 20th century. Rosebay rhododendron is projected to increase dramatically in coming years as its canopy associate, eastern hemlock (*Tsuga canadensis*), succumbs to the ravages of hemlock woolly adelgid (*Adelges tsugae*). In this way, rosebay rhododendron will once again demonstrate its versatility as nature's band-aid, sheltering cool, fast-moving mountain streams from overheating by absorption of solar radiation.

trail passes through a montane oak forest with Virginia pine (*Pinus virginiana*), white oak, and chestnut oak in the canopy. This is well above the expected elevation limit for Virginia pine, but since this is the most southerly of our hikes, the climate at this elevation may be more forgiving. This stretch of trail does not have enough pine to warrant calling the natural community a montane pine forest and woodland, but the story might be different if fire were reintroduced here. Without fire, the oaks predominate. The trail alternates between ridgetops

Small-statured northern red oak trees with their trademark "ski trails" are well-spaced toward the summit of Blood Mountain on the Appalachian Trail.

The tree canopy is thin at the summit of Blood Mountain, the highest point on the Appalachian Trail in Georgia. This Appalachian Trail shelter has the distinction of being on the National Register of Historic Places.

and open slopes, and there are several rocky areas where pines are more prevalent. Given the rocky soils, it is a bit surprising that there is a nice array of wildflowers tucked into spaces around the rocks.

As you ascend farther, you'll pass more areas of exposed rock, and the canopy becomes more open. You'll also notice that the trees (mostly northern red oak) become shorter and more gnarled, and some tree species that cannot endure the exposure, harsh winter temperatures, and thinner soils (such as tulip-tree) disappear.

Before you reach the summit, the natural community changes yet again as you ascend a south/southeast-facing ridgeline, where there are open rock outcrops. Most noticeably, the forest structure is different. Wide swaths of bare rock are interspersed with tall stands of Catawba rhododendron, with gnarled northern red oaks and small-statured serviceberry (*Amelanchier* sp.), no taller than 15 ft, scattered within the clumps. This stretch might make you question whether you're even in a forest. Here you'll also start getting nice views to the left (southwest) through some of the openings.

Soon you'll reach the two-room Blood Mountain shelter, built in 1934 by the CCC and on the National Register of Historic Places. It was restored in 2011 by volunteers from the Georgia Appalachian Trail Club, who replaced most of the wood as well as the roof. You can climb the rock beside the shelter to enjoy the view or sit inside in wet weather.

On your descent, pause in several places and estimate the percentage cover of the canopy—by this we mean how much of the canopy would cast shade over the ground. Even in dense forests, this is rarely 100%. Because subtle differences are harder to discern, ecologists often use a sliding scale to estimate percentage cover: 0–1%, 1–5%, 5–10%, 10–25%, 25–50%, 50–75%, 75–95%, 95–99%, 99–100%. Why is this measurement of interest? Changes in canopy cover may help us interpret patterns in the forest structure—for example, a lower cover in the canopy in the outcrop area permits better development of the shrub layer. You may also notice effects of canopy cover on the herb layer as you descend.

If it's a clear day, you'll have views to the south as you descend along the ridge. The trail curves back to the left and descends the series of switchbacks as you return to the taller and denser canopy of the montane oak forest. Make sure you turn left from the Appalachian Trail onto the Byron Herbert Reece Memorial Trail, 1.5 mi from the summit and 3.7 mi into your round-trip, to return to the parking area.

While you're in the area, you owe it to yourself to visit the Walasi-Yi Interpretive Center, a half mile south of the Byron Herbert Reece parking area at Neels Gap. It is an oasis for hikers who are about 30 mi out from the start of the Appalachian Trail and have descended from the highest point of the Appalachian Trail in Georgia—Blood Mountain, which you just summited. Some have abandoned their boots in disgust; you'll see hundreds of pairs dangling from the large tree near the breezeway where the Appalachian Trail passes through. The Walasi-Yi Interpretive Center was constructed by the CCC in the 1930s; it caters specifically to the needs of through-hikers who hope to reach Mount Katahdin in Maine, which is 2,154 mi away. And it is of interest as the only place on the 2,175 mi of the Appalachian Trail where the trail passes through a man-made structure.

2. SOSEBEE COVE

Why Are We Here?

Our ecological goal is to appreciate the value of "pocket parks," those natural-area gems that are essential components of any diversified conservation portfolio. We'll also consider the ecological problems created by forest fragmentation and the reasons why it is also important to conserve large, unfragmented tracts.

Just 175 acres in Chattahoochee National Forest, Sosebee Cove is an easily accessible rich cove forest. The diversity of gorgeous wildflowers carpeting the forest floor may convince you that this is a pristine landscape. However, you'll count fewer species of trees in the canopy than you might expect. There are other signs that this little treasure is not untouched wilderness. For one thing, there are a number of large trees that are of the same diameter and presumably of similar age; an even-aged canopy is characteristic of a second-growth forest. Moreover, a power line easement bisects the property, providing a

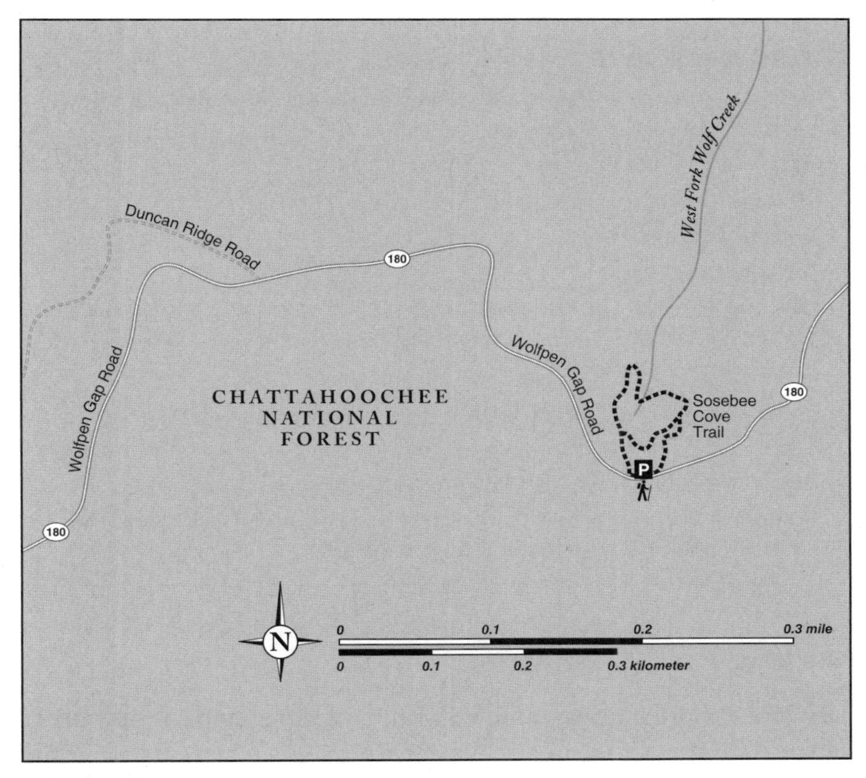

2. Sosebee Cove

corridor for the infiltration of exotic invasive species. Although we might wish that this were an old-growth forest without a power line right-of-way, rich jewels like Sosebee Cove, while not pristine, do contribute much to the conservation of forests. Despite its history of logging in 1903, the small preserve remains astoundingly rich in species. In your exploration of the two nested loop trails, we'll ask you to think about the conservation value of such places, beyond the obvious delight of all who appreciate the beauty of wildflowers.

You can thank former forest ranger Arthur Woody (1911–45) for this natural garden; the trail was dedicated in his memory in 1958. One of the first two US Forest Service foresters in the Chattahoochee National Forest, Woody spent his career restoring wildlife and fish populations here, as well as purchasing additional tracts of land for the national forest. During his career he added 250,000 acres of land to the Chattahoochee's holdings, including Sosebee Cove.

2. SOSEBEE COVE

Highlights: A pocket of north-facing cove forest, richly decorated with a stunning diversity of wildflowers

Natural Communities: Cove forest (rich subtype)

Elevation: 2,200 ft

Distance: 0.3 mi, two loops

Difficulty: Easy

Directions: From Blairsville, GA, take US 19/129 south about 9.5 mi. Turn right, heading west, onto GA 180. Go 2 mi past Vogel State Park to Sosebee Cove parking area on the right, along the road.

From Dahlonega, GA, take US 19/129 north for 23 mi. Turn west (left) on GA 180. Go 2 mi past Vogel State Park to Sosebee Cove parking area on the right, along the road.

GPS: Lat. 34° 45′ 44″ N; Long. 83° 56′ 51″ W (trailhead)

Information: U.S. Forest Service, Chattahoochee-Oconee National Forests, Brasstown Ranger District, 1881 Highway 515, P.O. Box 9, Blairsville, GA 30514, (706) 745-6928

The Hike

Descend a short stairway, and you will have the option to walk either left or right to start the upper loop. There are green blazes to mark the trail, but you'll always be within sight of the road, so you can explore in any direction without fear of getting lost. Turn left and walk the top of the trail parallel to the road. You can't miss a giant yellow buckeye (*Aesculus flava*), second in size only to the Georgia state champion, on the side of the trail just below the road. Over 15 ft in circumference, this huge tree requires that you join hands with two others to reach around its massive girth.

Follow the trail as it curves to the right and descends past the power line easement that cuts through the property. Easements are common here in the East, but cuts like this have an especially large impact on preserves like Sosebee Cove by breaking an already small tract into even smaller pieces, a process called fragmentation. The negative consequences of forest fragmentation are discussed in the sidebar.

After crossing the power line easement, you'll soon come to an intersection, where you can either go right to cross the preserve or continue straight on the trail to eventually reach the back of the tract. If you continue straight, you'll reach the bottom of the bowl-shaped cove, where the trail then turns right to cross Wolf Creek before you reach signs that tell you what you have already discovered: Sosebee Cove is a "Botanist's Paradise." On the signs, canopy trees are grouped and listed at the top, followed by smaller trees in the understory, followed by a few of the many wildflowers that grace the trails.

Near this point, the trail connects to the upper loop and cuts back toward the road. Instead of going straight back to the road, turn right to walk through the heart of the preserve. You'll pass through a stand of large but similarly sized, arrow-straight tulip-trees (*Liriodendron tulipifera*) as you approach the wooden bridge that crosses Wolf Creek. You might wonder what it was like here when all of these trees germinated at the same time. Is it possible that they were planted?

History tells us that this is not a plantation but that Sosebee Cove was logged in 1903, so these large tulip-trees are now well over 100 years old. They established themselves in the cut-over site on their own. Tulip-tree has several characteristics that make it successful here. Unlike most tree species that are either slow-growing and long-lived or fast-growing and short-lived, tulip-tree is both fast-growing and long-lived. At lower elevations in the mountains (below about 2,500 ft), tulip-tree is the primary pioneer species on cut-over lands. A moist, open site is perfect for its establishment, and when you see even-aged stands of nearly pure tulip-tree in the mountains, you can guess that the forest regenerated after a clear-cut. The lightweight, winged seeds helicopter far and wide, so we can surmise that tulip-tree readily seeded here and germinated in near-perfect conditions after logging. You may be interested to learn more about even-aged forest stands in our "Second-Growth Forests" sidebar in hike 12 (Whiteside Mountain).

Wildflowers carpet the feet of the forest giants in Sosebee Cove. In early spring, look for several species of trillium (easy to spot with their whorls of three large, leaflike bracts): sweet white trillium (*Trillium simile*) (white with dark centers); large white trillium (*T. grandiflorum*); and the amusingly named purple toadshade (*T. cuneatum*), with deep maroon flowers perched directly on top of the leafy bracts. Another sessile trillium found here is yellow trillium or yellow toadshade

Forest Fragmentation

Across the planet, forests have become increasingly fragmented by human activities, and forest fragmentation has been happening in the southern Appalachians since the arrival of humans on the scene many millenia ago. Once-continuous tracts of forest are now splintered into a mosaic of smaller patches, isolated from one another by areas of nonforested habitat, including recently logged areas, single-species plantations, pastures, agricultural fields, and urban or industrial development. The resulting fragments have much more "edge" habitat than the previous continuous forest, meaning that there is more interface between open, disturbed areas and the adjacent forest. Increase in edge habitat occurs at the expense of forest interior habitat, which is important for the survival of many species.

In the case of fragmented forested tracts like Sosebee Cove, the smaller areas of interior forest and their proximity to forest edges mean that some forest interior birds are more susceptible to nest parasites, such as the native brown-headed cowbird (*Molothrus ater*). These deadbeat parents lay their eggs in the nests of other birds, and the larger cowbird nestlings easily outcompete their step-siblings for food and care from their adoptive parents. While not the most ethical or responsible move for a parent, it's a win for the cowbirds in the game of natural selection: their offspring are raised for "free," without the energy cost of building a nest or providing for young. This enables the parasitic cowbirds to produce even more young with other foster parents, at the expense of the foster parents' biological children. Nest parasitism by cowbirds is reduced in the interior of large, unfragmented forest tracts because the cowbird, originally a prairie species, is reluctant to venture very far from a forest edge.

Even in the absence of overt fragmentation, highway and power line corridors create problems for native species. Such corridors are often superhighways for invasive exotic plant species, which are nonnative species that thrive and take over their adopted homes. You'll see at least two invasive plant species from Southeast Asia in Sosebee Cove, kudzu (*Pueraria lobata*) and tree-of-heaven (*Ailanthus altissima*), both of which have taken advantage of the open canopy and disturbed soil to invade the preserve. These species are fast-growing, and preserve managers keep a wary eye on such aggressive aliens. Because these two species also require open areas with plenty of sunlight, the rest of the cove is safe, at least for now. But if a storm topples one of the large canopy tulip-trees (*Liriodendron tulipifera*) and opens up space, these invaders are poised for a hostile takeover.

Another problem with easements is that they form barriers to movements of animals that are averse to traveling through open areas. Restricted movement reduces the connectivity important for maintenance of healthy animal populations; the services that animals provide to plants, such as pollination and seed dispersal, may be similarly curtailed. Maintenance of corridors for wildlife movement between forested patches is a prime consideration for landscape managers, and novel techniques for creating corridors for movement include construction of tunnels and forested overpasses, and planning for "green" corridors along natural travel pathways such as streams and rivers. Planning for unimpeded movements of both animals and plants is taking on a new urgency as we recognize the importance of animal and plant migrations in the face of global climate change.

Sweet white trillium grows in abundance in bowl-shaped Sosebee Cove, nestled beneath a canopy of even-aged tulip-tree.

(*T. luteum*), which is like purple toadshade only with yellow flowers. White spikes of foamflower (*Tiarella cordifolia*) and showy orchid (*Galearis spectabilis*) can also be seen in April. When you get to the wooden bridge crossing the creek, look downstream and notice that a different set of species favor having their feet wet, notably summer-flowering herbs that include umbrella leaf (*Diphylleia cymosa*) and jewelweed (*Impatiens capensis*).

You can continue along the trail, turning left when you reach the next junction to climb back out of the cove, or you can linger to admire the spectacular natural garden. When you reach the stairs again, the only section of trail you have not walked lies straight ahead—be sure to at least peek at one of the largest tulip-trees in the cove, well over 16 ft in circumference and clearly of an older generation than those that germinated in the early days of the automobile. Perhaps it was the seed from this tree that spawned the second-growth forest below.

Sosebee Cove, given its small size, remains vulnerable to influence and harm from outside its boundaries. Yet small pockets of nature like this one protect our natural heritage and put natural inspiration easily within reach.

3. BRASSTOWN BALD

Why Are We Here?

Brasstown Bald, the highest peak in Georgia at 4,370 ft, has surprisingly few visitors despite its lovely trails, accessibility for those with physical disabilities, and spectacular 360° views from the platform at the summit. The visitor center, also at the summit, provides information on both natural and human history as well as a row of rocking chairs; from here you can take in the views and rest your legs. A recent visit over Memorial Day weekend yielded an incredible floral display of both mountain laurel (*Kalmia latifolia*) and Catawba rhododendron (*Rhododendron catawbiense*). Because of the high elevation, spring arrives late at Brasstown Bald, and the season of early spring wildflowers is compressed into a few short but glorious weeks.

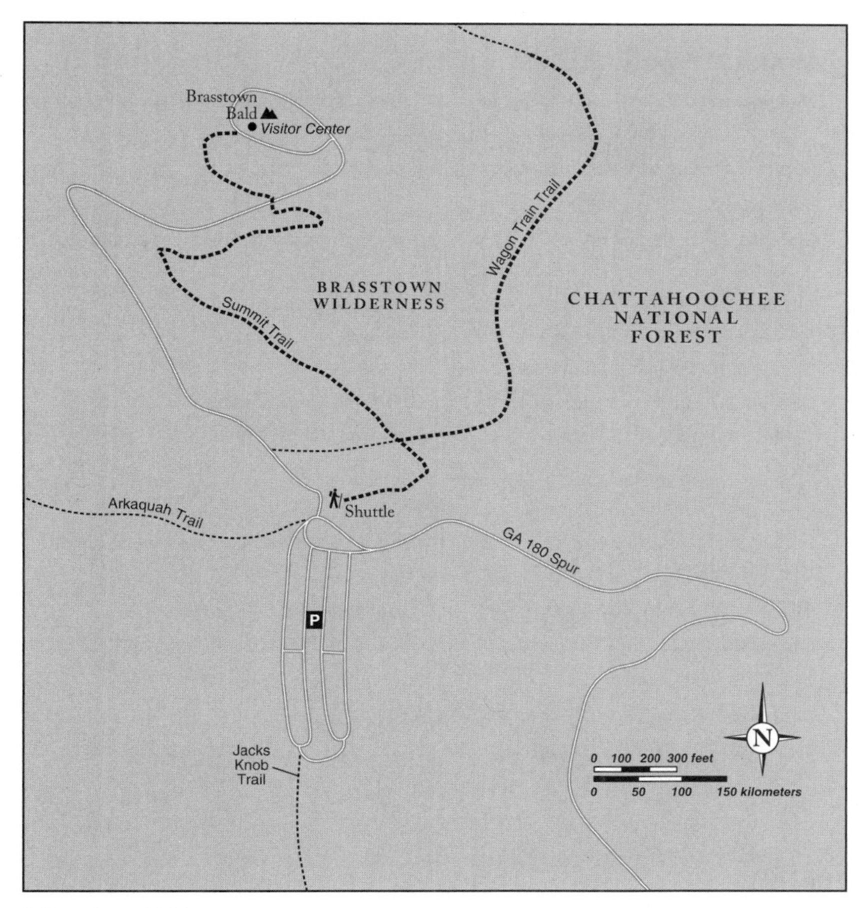

3. Brasstown Bald

You might be surprised to discover that, here in the southernmost portion of the southern Appalachians, the highest summits do not have spruce-fir forests. Instead, the summit is cloaked by a montane oak forest and woodland. Make no mistake – you are well above 4,000 ft, and the mountaintop reflects the extreme conditions you'll encounter at any high-elevation summit; but this peak has attributes that reveal its southern hospitality, too. We'll compare the more exposed summit forest with one that is protected and situated below the summit. You can make this comparison fairly easily by hiking down from the top of the short Summit Trail, then turning right just before you reach the parking lot to hike the first section of the Wagon Train Trail.

3. BRASSTOWN BALD

Highlights: Highest peak in Georgia, with 360° views; interesting comparison between two higher-elevation forests, a montane oak forest and a northern hardwood forest

Natural Communities: Montane oak forest, northern hardwood forest

Elevation: 4,784 ft (summit)

Distance: 1.2-mi round-trip on the Summit Trail plus 1-mi round-trip on the first portion of the Wagon Train Trail for 2.2 mi total

Difficulty: Steep (500 ft) climb on paved trail to summit; Wagon Train Trail is easy walking. There is also a fee-operated tram that will take visitors to the summit and a visitor center that operates seasonally. There is an elevator to the viewing tower.

Directions: From the intersection of US 129 with GA 180, drive north on GA 180. Turn left on the GA 180 Spur and drive 3 mi to the parking area for Brasstown Bald.

From the intersection of GA 17/75 with GA 180, drive west on GA 180. Turn right on the GA 180 Spur and drive 3 mi to the parking area for Brasstown Bald (small parking fee, restrooms, picnic area; shuttle, visitor center, and concession stand open seasonally).

GPS: Lat. 34° 52′ 20.5″ N; Long. 83° 48′ 64.3″ W (parking area)

Information: Chattahoochee-Oconee National Forests, 1755 Cleveland Highway, Gainesville, GA 30501, (770) 297-3000

Brasstown Bald Visitor Center (706) 896-2556

The Hike

The green-blazed trail to the summit of Brasstown Bald begins behind the concessioner's building. Although short and paved, it is quite steep. You could opt to take the shuttle to the top, enjoy the views, and then take your time on the easy walk back down to pause and appreciate your surroundings. Because this seems to be a popular option for many visitors, this hike will start from the summit and descend to the intersection with the Wagon Train Trail. If you decide to hike to the top, there are many benches along the way to take a breather.

Start at the observation tower. From your bird's-eye view from the tower, one of the first things you'll notice at the summit is that it is densely vegetated rather than exposed, rocky, or grassy. Although some Cherokee legends claim that the summit was once cleared for farming, the term "bald" in this case does not mean that it is an open, treeless expanse.

Depending on the time of year, you may notice that the canopy is mostly deciduous, though there is a small component of planted red spruce (*Picea rubens*) and evergreen shrubs that add their greenery to the winter landscape. If you visit in late May to early June, you'll catch the leafing-out of the hardwood trees punctuated by deep magenta blooms of Catawba rhododendron and the pale pink of mountain laurel. It's a bit like a study in pointillism, with tiny dots in many shades of green interspersed with larger blotches of pale pink and deep magenta flowers.

Cross the road and locate the green-blazed trail heading down to the parking area. You've seen the forest canopy from the top, and now you'll enter the forest. It might seem surprisingly short in stature—the tallest trees are bent and gnarled, and their height is only about 15 ft near the summit, though they get taller as you descend farther from the exposed summit. Most of the twisted trunks are oaks, with northern red oak (*Quercus rubra*) dominant and some white oak (*Q. alba*) mixed in.

At 4,700+ ft, the summit of Brasstown Bald could plausibly have red spruce and more yellow birch (*Betula alleghaniensis*), but although there are some planted spruce at the base and summit, there are no recorded native populations of red spruce in Georgia. Why not? The elevation and other environmental factors here are right for red spruce. The planted red spruce at the base and summit are thriving. Why is this summit not a red spruce forest? Clues to the answer may lie in the distant past, as discussed in this hike's sidebar.

Cross the service road that leads to the summit and continue to descend the trail, noting other species that share the summit with the red oaks. You will see two species of maples: mountain maple (*Acer spicatum*), with its coarse-veined leaves, and the more ubiquitous sugar maple (*A. saccharum*). The pink blooms of Catawba rhododendron are replaced by the white flowers of rosebay rhododendron (*Rhododendron maximum*) on the lower slopes.

You will soon reach the well-marked intersection with the Wagon Train Trail, which was built in the 1950s as part of Route 66 but was never completed. The parking area is about 100 yards ahead, but turn

The Case of the Missing Spruce

High-elevation species, such as red spruce (*Picea rubens*), are absent from some of the high peaks in the southern Appalachians that might be expected to support them. It's probable that red spruce and even Fraser fir (*Abies fraseri*) once covered the summits of places like Brasstown Bald, the highest point in Georgia at 4,784 ft. Although glaciers never reached the southern Appalachians, the climate was much cooler during the last Ice Age. As a result, it seems certain that the Brasstown Bald summit, along with others of similar elevation, supported a much different forest. We can't be certain that a boreal spruce-fir forest was here—the summits might even have been above treeline—but it seems very likely.

Imagine yourself once again on the tower at the summit of Brasstown Bald, looking down on the surrounding mountains. If the valleys were suddenly flooded to 4,000 ft of elevation, what you would see rising above the water would be little islands of mountain summits. You'd be on the highest island in Georgia, but you would see a few islands nearby, Blood Mountain to the south (on a clear day) and Rabun Bald and nearby peaks to the east.

Instead of islands separated by water, though, what you're really seeing are islands of cooler climate. These islands of cooler climate shrink as the climate warms, and as it cools, they expand and other peaks may surface. During the last Ice Age, these cooler island peaks were likely covered in boreal forest. As the climate warmed, the sea of temperate forest (probably northern hardwood and montane oak forests) filled the valleys and climbed the slopes, eventually covering the tops of all the Georgia mountains. The boreal forest, its key species unable to tolerate the warmer temperatures, disappeared first in the warmer valleys and then from the summits of the southernmost portion of the southern Appalachians.

Red spruce persisted to the north, from its southernmost population in the Smokies and Cowees to the higher mountain ranges like the Black and Craggy Mountains in North Carolina, thence north to New England. The combination of higher elevations and more northern latitudes allowed red spruce to persist.

The temperature has fluctuated greatly in the last 18,000 years, with periods of moderate warming and cooling. In today's climate, red spruce could grow and thrive above 4,000 ft at Brasstown Bald and other Georgia mountain summits. But how could red spruce get back there? The seed source has disappeared. The Cowee Mountains, which are 50 mi to the northeast, have the nearest natural stand of red spruce. Therefore, the high mountains of Georgia, while climatically similar to mountains farther north, have a different natural community.

From this observation, we learn an important lesson of ecology: not every species will be found where conditions are favorable for its success. Natural barriers to dispersal and, in the case of Brasstown Bald, the vagaries of time and history can conspire to prevent some species, like red spruce, from occurring where they might be otherwise expected.

left to follow the Wagon Train Trail. This trail stretches nearly 8 mi to its end at the town of Young Harris, but you will only walk a short distance, a half mile or so, to see a very different forest that contains features that reflect events during the Ice Age and thereafter.

The old road narrows, goes through a gate, and passes through a short stretch of private property before crossing the boundary into the Brasstown Wilderness, part of the Chattahoochee National Forest. You'll enter a tunnel of rosebay rhododendron before emerging, about a quarter mile from where you started, into a more open area of old road. You've journeyed around from the southern slope of Brasstown Bald and are now walking along the northeast-facing slope. As the trail opens up, look to your left to see a jumble of giant boulders stretching up the slope. The jumbled rocks extend down the slope to the right of the trail as well.

What you see is a boulderfield, also an artifact of the past. Following the Ice Age, the climate warmed, but not gradually. Evidence of the period 18,000 to 12,000 years before the present suggests that the earth's climate oscillated wildly, with alternating periods of rapid warming and rapid cooling. Repeated cycles of freezing and thawing broke apart the north-facing slopes on many summits, causing massive landslides and resulting in boulderfields like the one you see here. It's a hazardous environment, one where dense leaves and thick moss can obscure loose stones and the deep cracks between them, so see what you can from the trail. (The moss mats and other vegetation on the rocks are also fragile, and walking on them can do considerable damage.)

Once you have observed the tumbled rocks and physical environment of the boulderfield, notice that the forest that grows here is quite different from the forest you observed on the Summit Trail. Rather than oaks, the dominant trees here are yellow birch, which have peeling, shiny bark. This forest is one subtype of a northern hardwood forest called a birch boulderfield. Rather than the usual diversity of canopy species that inhabit a northern hardwood forest, yellow birch dominates.

Why does yellow birch thrive here at the expense of other species? You are at nearly the same elevation as you were on the Summit Trail, so some other reason is needed to explain the shift in species composition. It turns out that yellow birch is adept at exploiting these inhospitable environments. It can establish itself and grow in the very shallow soil (mossy mats, really) on top of the boulders, and its roots can then creep around the rocks and extend into the soil. This gives yellow birch a distinct advantage over other species in these rocky environments. Yellow birch is a remarkably versatile species, and you can learn more about its adaptations in the "Yellow Birch—Engineered for Disaster" sidebar in hike 17 (Alum Cave).

Unstable boulderfields continue to shift under the influence of gravity, wiping out stands of yellow birch that are then replaced by a new forest. You can see this by noting that yellow birch trees in a given area are often about the same size, which indicates that they were all established around the same time following the last disturbance.

Walk to the right side of the trail and peer at the boulderfield below. Although most trees have difficulty rooting in these places, spring wildflowers thrive in the small but sometimes deep pockets of soil. Because of the elevation, the spring blooming season is compressed, so you might see the earliest wildflowers, such as trout lily (*Erythronium umbillicatum*), blooming at the same time as later flowers like Dutchman's breeches (*Dicentra cucullaria*) and red trillium (*Trillium erectum*).

Unlike the birch boulderfield, the summit is covered with northern red oak. Heavy red oak acorns would either fall through the spaces in the rocks or land on top of the boulders, where there's not enough soil for germination and establishment. Although oak seedlings can tolerate some shade, they can't compete here under fast-growing birch. However, their slow growth and gnarled form is well-suited to the

Even-aged yellow birches dominate the boulderfields like those found on the first section of the Wagon Train Trail. These natural communities formed when freeze/thaw cycles following the last Ice Age broke apart rocky, north-facing slopes at high elevations.

Red trillium takes root in the pockets of rich, moist soil found in the spaces between loose boulders.

extreme exposures found on the summit, and red oak can regenerate in its own shade. Fast-growing yellow birch, on the other hand, is likely to get blown over on the exposed environment of the summit. It is also likely that the summit habitat is drier, so oaks as well as drought-tolerant herbaceous plants have an advantage. You'll find more diversity in the herb layer in the moist pockets of deeper soil on the north-facing boulderfield, where summer temperatures stay cool.

When you have finished exploring the boulderfield, you have the option to walk farther down the Wagon Train Trail or go back the way you came, turning left at the intersection with the Summit Trail to return to the parking lot. At the trailhead, note the two healthy red spruce that were planted there and think about how this mountain must have looked covered in an evergreen canopy during a cooler climate 18,000 years ago.

4. TENNESSEE ROCK TRAIL

Why Are We Here?

If you were taking the class called Forest Ecosystems of the Southern Appalachian Mountains at the Highlands Biological Station, the Tennessee Rock Trail would be the ideal location for your final exam. After two intensive weeks of class, you would be a seasoned forest ecologist with a keen ability to interpret forested landscapes. You would hike the 2.2-mi loop trail and then answer a single question: What's the story here? (Note: This is an idea we have never actually tried with our students; we could never get past our inherent concern that students would find this a terrifying experience.)

However, you are not taking our class (yes, you're entitled to a sigh of relief). Nonetheless, our goal is to interpret the forests here. There are so many interesting ecological stories at Black Rock Mountain State Park that it is hard to choose just one. Some stories are based around various environmental factors that play roles in dictating the type of forests found here. Other stories can only be understood by

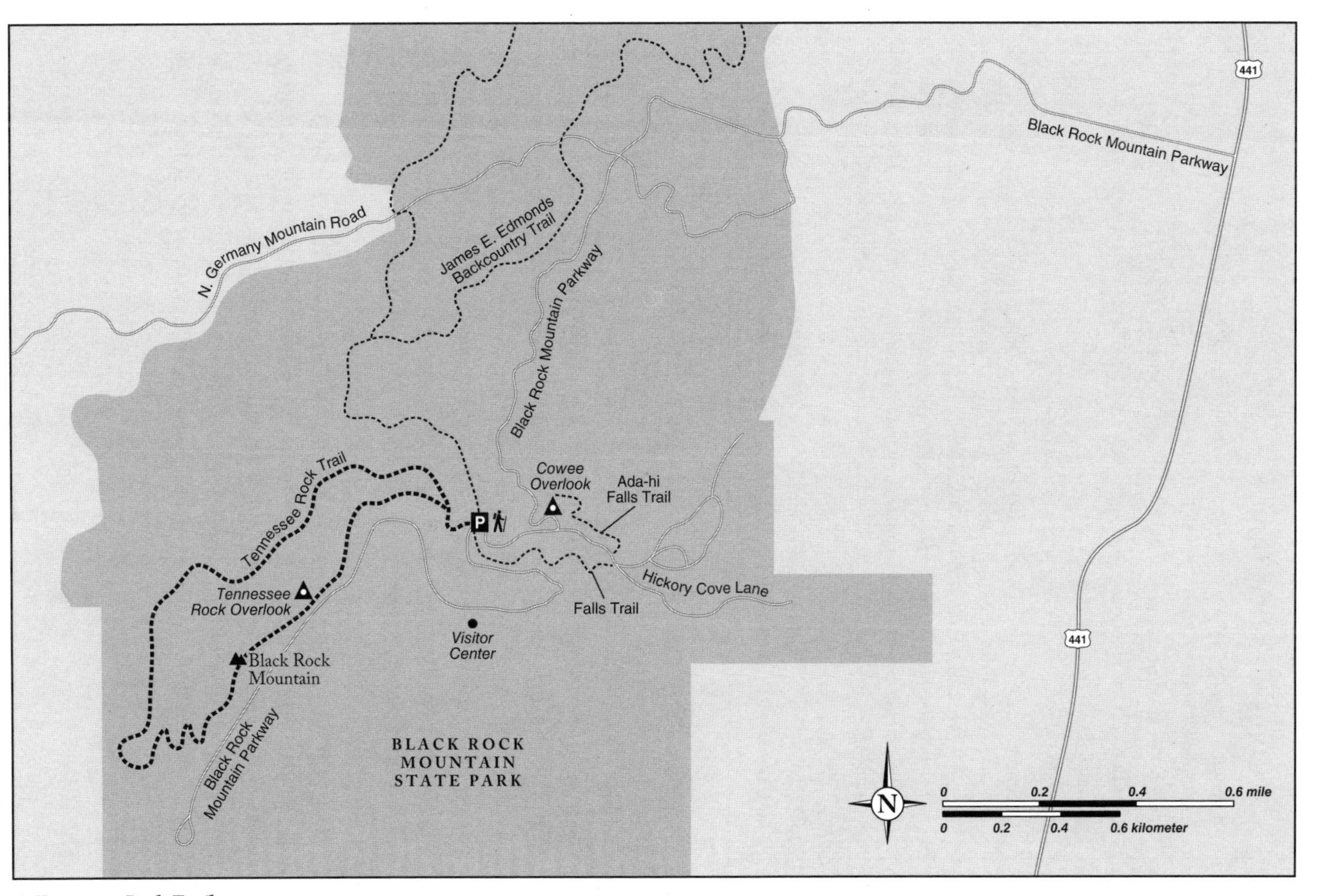

4. Tennessee Rock Trail

4. TENNESSEE ROCK TRAIL

Highlights: Amazing variety of habitats in a lovely and accessible loop trail, plus a close-up view of recent tornado destruction and recovery

Natural Communities: Cove forest (rich subtype), northern hardwood forest, oak forest

Elevation: 3,250 ft (Tennessee Rock trailhead); 3,640 ft (Black Rock Mountain summit)

Distance: 2.2 mi

Difficulty: Mostly easy walking with a few moderate climbs

Directions: Black Rock Mountain State Park is between Dillard and Clayton, GA, off US 23/441. Look for small brown signs as you approach Mountain City. Turn west (left if you are coming from Clayton) on Black Rock Mountain Rd. The road climbs steeply—follow signs for the park. At the T, turn right to follow signs for the visitor center. The parking area and trailhead are on the right before you reach the visitor center (small parking fee, self-service kiosk, picnic table, but no restrooms, which are at the visitor center).

GPS: Lat. 34° 54′ 25.5″ N; Long. 83° 24′ 43.3″ W (trailhead)

Information: Black Rock Mountain State Park, Mountain City, GA 30562, (706) 746-2141

knowing the history of the site, which stretches thousands of years into the past. An ecological story can never be explained by a single environmental factor or one past event, and that complexity makes the story interesting but difficult to piece together.

First, let's briefly consider the spatial and temporal parts of our journey. The trail encircles the summit of Black Rock Mountain, climbs the summit ridge, crosses the summit, and finally descends to the parking area. We will also travel deep into the past and then back to the present. The story spans thousands of years, with change as the only constant.

If you stop by the visitor center before your hike, you can purchase an excellent and informative interpretive guide written by Anthony Lapros and Dustin Warner, with natural history information corresponding to numbered signposts along the trail. The booklet is beautifully illustrated and offers additional discussions of ecology and wildlife.

The Hike

From the parking area, you have two options: the 7.2-mi James Edmonds Trail or the 2.2-mi Tennessee Rock Trail. Follow the yellow blazes and signs for the Tennessee Rock Trail and ascend until you reach the intersection where the loop hike begins. You'll follow the loop counterclockwise, so bear right at this intersection to start your hike, continuing to follow the yellow blazes.

The trail begins in an oak forest, a natural community that is ubiquitous in the southern Appalachians and one that you'll see all along your pathway here. However, you will see subtle changes in the species and forest structure as you circumnavigate the summit of Black Rock Mountain.

The forest you are hiking through now has a well-developed structure that results from more than 100 years without major disturbance. If you were to take a vertical cross section of this forest, you would discover a canopy of oaks and hickories, an understory of smaller trees below the canopy, a sapling and shrub layer, and finally an herb layer.

The oak forest seen along most of this hike is different from a montane oak forest because the elevation is lower and the climate is less demanding, so the forest trees are able to grow straight and tall. Northern red oak (*Quercus rubra*) and white oak (*Q. alba*) are canopy dominants, along with tulip-tree (*Liriodendron tulipifera*). The understory is fairly rich, and throughout the spring and summer, you'll be walking through a lovely woodland garden rich with wildflowers.

As the trail curves to the left and the slope becomes north-facing, you might notice subtle differences in the canopy. At lower elevations, the direction of the slope is important because it determines the amount of solar energy received. South-facing slopes are hot and dry in the summer. North-facing slopes remain cool and moist, even in the South, and these often harbor species that are most commonly found farther north.

The trend toward increasingly cool and moist conditions on northerly slopes is most evident in the herb layer, so watch closely as you follow the trail around to the left. As you near a small seep, you'll notice that the herb layer becomes lush, with wildflowers in multiple layers. The diversity, or number of species, increases, as does the sheer volume of plants. While moisture surely plays a major role in this story, the soil plays an important supporting role. Black Rock Mountain is underlain by biotite gneiss, and the soil that develops is rich in minerals, allowing for a greater variety of wildflowers.

You will soon see a sign where you can follow a path to the left, a short spur trail rich with wildflowers, to a natural amphitheater with steep rocks forming a north-facing wall. The wall is not solid, however. Rather, it is a pile of boulders, so please heed the signs and do not venture beyond the barriers to climb this wall. It may look solid, but the rocks can shift unexpectedly, and there are big gaps between them. In addition, the wildflowers and mosses that form a living mantle atop these rocks are quite fragile. What you see in front of you is actually a natural feature called a boulderfield, and this part of the Black Rock Mountain story stretches far into the past.

In the southern Appalachians, boulderfields formed between 12,000 and 18,000 years ago as the earth's climate fluctuated. The periods of relatively rapid freezing and deep ice followed by thawing broke apart solid rock, resulting in massive rockslides. Boulders tumbled into the valleys and formed deep boulderfields. The resulting rocky landscape poses challenges for hikers as well as for the trees attempting to take root here.

Yellow birch (*Betula alleghaniensis*) is the most adaptable to these conditions, germinating in the moss and then growing roots around the rocks and into the soil, though here you will also see quite a few yellow buckeye (*Aesculus flava*), which are also successful in rocky habitats. The difficulties experienced by trees, however, make for a more open canopy, with more sunlight reaching the forest floor and the carpet of wildflowers and other herbaceous plants.

Retrace your steps on the spur trail and turn left when you reach the main trail. You'll continue on the rich, northeast-facing shoulder of the mountain, through a diverse oak forest with hickories (*Carya* spp.) and tulip-tree. The trail curves right and, shortly thereafter, makes a left turn. Here the trail joins and follows the path of an old roadbed, descending gradually and continuing southwest before turning due south.

As you follow the road, you'll enter a much denser, darker forest. Take a look around you and notice that the trees are mostly white pine (*Pinus strobus*), and their dense evergreen boughs keep the forest cool and quiet. Why are there pines here and not elsewhere? Why do they grow so close together? We'll return to these questions after a brief digression.

Something rather interesting about white pines is that their age is easy to determine. For every year of growth, white pines produce a single whorl of branches, consisting usually of 3 or 4 branches arranged like the spokes of a wheel. This behavior is unlike that of most of our southern pines (all in the yellow pine group), which produce multiple whorls of branches each year; the number of whorls varies depending on site conditions and competition. Unlike the yellow pines, white pine tends to retain its lower branches. Even where these branches have been shed, you can still usually see the points of attachment. You can thus get a rough estimate of the tree's age by counting the whorls of branches.

Look carefully at the stand of pines from different angles. Two points will become clear. First, this stand was not planted. If it had been, you'd be able to discern neat rows of trees. Second, these trees are nonetheless the same age. Indeed, before these pines were here, this part of the mountain was open; where you are standing was an open agricultural field until the 1950s.

Pine seeds are lightweight, are produced in abundance, travel far, and readily germinate where there is sunlight and bare mineral soil. Seedlings can tolerate some level of drought as they grow rapidly in the race to the canopy. Abandoned agricultural fields present an ideal venue for pine germination and establishment, provided there is a seed source nearby. In the Piedmont, yellow pines predominate as the "old-field" pines, but in the mountains white pine plays this role, along with tulip-tree.

The ability to seed quickly and grow rapidly to maturity has a trade-off: pines are, relatively speaking, short-lived. Pines are the James Deans, Kurt Cobains, and Amy Winehouses of the tree world—they live fast and die young. In the mountains, they are susceptible to winter storms that bring ice and wind damage; their heavy evergreen limbs are easily blown down or encumbered by heavy snow and ice. In addition, they are unable to replace themselves in an aging stand. Their requirement for nearly full sunlight for establishment means that they are unable to grow in the subcanopy. You may see trees that are losing their fight for canopy dominance. Pines that don't win space in the canopy will eventually die, a process that thins the stand naturally and provides the remaining pines additional space to grow. Ecologists call this process "self-thinning" to distinguish it from thinning imposed by foresters.

You'll notice as you walk through this white pine stand that its density creates a simpler forest structure than that found in the oak forest on the north flank of Black Rock Mountain. If you made a list of the number of tree and other plant species here, you'd discover that the diversity has dropped significantly. Still, there are specialists in all kinds of forests. Pink lady slippers (*Cypripedium acaule*) thrive under pines, and if you visit in May, you might see these lovely orchids in flower.

At 1.7 mi, you'll still be in white pine forest as you travel south, but the stand includes tulip-tree, which may outlive the pine. The trail makes a sharp left and begins to climb a series of steep switchbacks. Take your time, for your journey through the next chapter of this forest's history is fascinating. What we'll describe (and what you'll see) is a case history of forest disturbance caused by a single destructive event: a tornado that passed through the park in 2011. We cover this case history in this hike's sidebar.

After traversing the tornado-damaged slope, the trail turns northeast after the switchbacks, climbing the ridgeline to the summit. Here, you are back in hardwood forest, but the thin soils of the ridgeline and increased exposure to the elements favor an open oak forest dominated by oaks (*Quercus* spp.) and hickories (*Carya* spp.). You'll see more wildflowers in the understory here as you make your way to the stone marker at the summit, 3,640 ft. The elevation is not that different from where you started, but the increased exposure to the elements along the ridgeline creates conditions that are clearly different. As one example, spring arrives a bit later up here.

The summit of Black Rock Mountain is covered by forest, but it marks a point on the Eastern Continental Divide and the intersections of three watersheds. The part of the park where you've just been walking—the west side—drains to the Tallulah River Basin, which eventually flows to the Mississippi. To the north, rainfall enters the Little Tennessee River basin, which also drains to the Mississippi. To the south and east, rainfall eventually makes its way into the Chattooga River and finally the Atlantic Ocean.

From here, you'll continue to descend the rocky ridgeline along the Tennessee Rock Trail until you reach the Tennessee Rock Overlook. Along the way the character of the forest again changes, with tall mountain laurel (*Kalmia latifolia*) and rosebay rhododendron (*Rhododendron maximum*) entering beneath the increasingly stunted canopy,

Tornadoes for Renewal

As you hike along the Tennessee Rock Trail in Black Rock Mountain State Park, you'll enter the path of a relatively recent tornado after you cross through the white pine (*Pinus strobus*) stand, around 1.7 mi from the trailhead. You'll immediately know that some destructive force has affected this forest. The canopy is more open here than anywhere else and is completely gone in places. Many trees are uprooted and lying on the ground. Other trees are simply snapped off at the top, leaving bare trunks as silent witnesses to this destructive force in the recent past. You'll notice that broken and downed trees alike are oriented in the same direction, pointing to the north.

The ravaged southwest flank of Black Rock Mountain will eventually recover from the 2011 tornado, but it will retain evidence of this violent event. The storm will become just another plot twist in the history of this mountain.

On April 27, 2011, a tornado roared over the western flank of Black Rock Mountain, leaving this destruction behind. According to park records, the tornado hit this section of the mountain with wind speeds of up to 165 mph. The forest had a large component of white pine, and the needled limbs added weight, and thus instability, in this tremendous windstorm. You can tell that this destruction is from a tornado, based on the relatively small area affected. Although you might see a few downed trees along other areas of the trail, the severe storm damage is concentrated here on the southwest flank of Black Rock Mountain.

You might look at the destruction and think sadly that this forest will never be the same again, and to some extent you would be correct. The only constant, really, is change; here it is just more dramatic. However, the big blowdown is not cause for despair, because the forest is already recovering through a natural healing process.

Look for signs of renewal. Downed and snapped-off trees provide important wildlife habitat, particularly for insects and, by extension, woodpeckers. You may see woodpecker holes in the standing dead trees. Coarse woody debris, as it is called by wildlife biologists, provides habitat for reptiles and amphibians. You might be able to make some predictions about how the next forest will look. As you climb the last switchbacks to the ridgeline, take time to look for tree seedlings poking up through the downed trunks. Are they mostly white pines, red maples (*Acer rubrum*), or tulip-trees (*Liriodendron tulipifera*)? Which of these species will become the next forest dominant?

It's fun to consider the future possibilities. If it is not too dry, tulip-tree may be the next dominant canopy species; it is fast-growing and long-lived, which give it good odds. However, white pine would also thrive in open sunlight and would tolerate drought better. Finally, red maple is so adaptable to a range of conditions that it's certainly within reason that it might eventually dominate. You'll find oak seedlings as well; these will grow slowly and inconspicuously at first, but in the end they may outlive whatever pioneering species establish themselves now. In a few places, you might have a hard time imagining a forest at all. Wild blackberry (*Rubus* spp.) has taken over large areas, and its dense, thorny branches and large leaves make it hard for any trees to establish beneath it. Nature's wheel of fortune has been spun by this tornado, and now we're waiting to see who will be the big winner.

as the ground beneath your feet grows rockier and the soil even thinner. Look for the forest to change again as you descend the ridgeline and enter a taller, more protected forest with sweet birch (*Betula lenta*), northern red oak, and white oak in the canopy.

Soon you'll reach an intersection and complete the loop hike. Turn right to take the spur trail that descends back to the parking area. The story of Black Rock Mountain hasn't ended, and the future of its diverse forests isn't clear. But you can bet that the mountain will continue to change, sometimes slowly as the forests grow and develop, sometimes quickly with a lightning strike or fierce storm. As ecologists, you can look for clues that help you predict the story's next plot twist, but the mysteries of the mountain are never fully revealed.

5. MARTIN CREEK FALLS

Why Are We Here?

In 1775, as the American colonies were gearing up for revolution and independence, naturalist William Bartram traversed this area far from the fray, documenting the flora, fauna, and various native American cultures. He detailed his journey in *Travels*, first published in 1791, inspiring English romantics with his flowery, descriptive prose. Today you will retrace his journey on part of the Georgia section of the Bartram Trail. Bartram's words describe Martin Creek Falls and a native tree, Fraser magnolia (*Magnolia fraseri*). You'll encounter two lovely cascades, Becky Branch and Martin Creek Falls, and we invite you to consider what the forest may have looked like on the cusp of the American Revolution.

If you have additional time to explore, spend some in the Warwoman Dell picnic area to learn about its rich history. The origins of the name Warwoman are hazy, but many stories seem to center around a female Cherokee leader who made major decisions for her tribe, such as going to war and fates of prisoners. More recently, this site hosted an attempt to build the Black Rock Railroad from Charleston, SC, to Cincinnati before the Civil War brought those plans to a halt. In the 1930s, the Civilian Conservation Corps (CCC) created a camp at this site, doing conservation work to restore areas damaged by logging.

The Hike

From the end of the parking area farthest from the road, walk back toward Warwoman Rd. On your left, you'll see several rectangular ce-

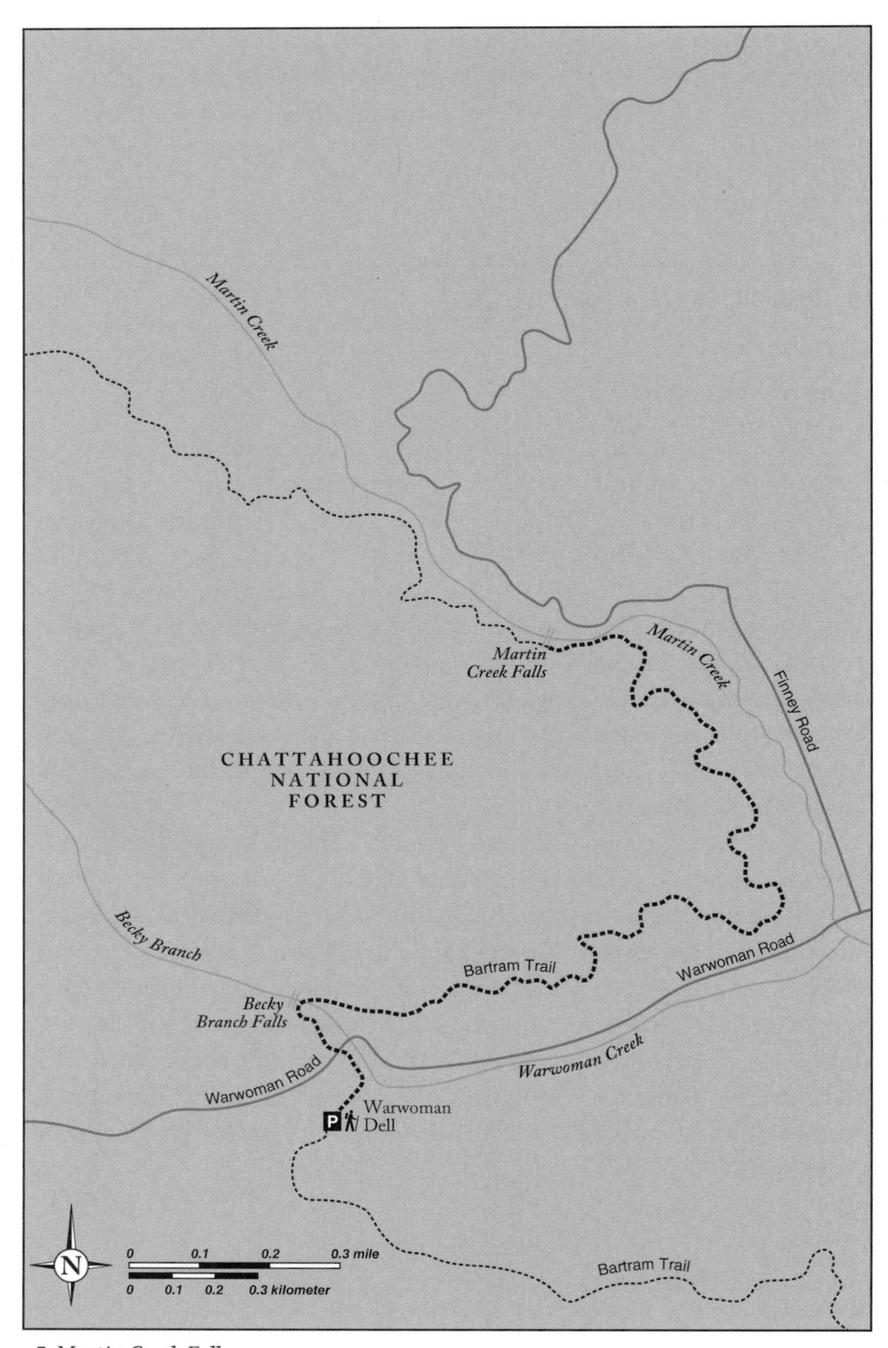

5. *Martin Creek Falls*

5. MARTIN CREEK FALLS

Highlights: Tracing the footsteps of explorer William Bartram, who first described Fraser magnolia, a southern Appalachian endemic, while perched below a cascade he called Falling Creek (Martin Creek Falls)

Natural Communities: Low-elevation pine forest, oak forest, cove forest (acidic sub-type)

Elevation: 2,000 ft (Warwoman Dell)

Distance: 3.6 mi round-trip

Difficulty: Easy to moderate

Directions: From Clayton, GA, head north on US 23/441 from the intersection with GA 76/2. Turn right on Warwoman Rd. (884E) and drive about 2.5 mi. The Warwoman Dell Recreation Area is on your right (picnic area with grills, drinking water, vault toilets).

GPS: Lat. 34° 52′ 55″ N; Long. 83° 21′ 4″ W (trailhead)

Information: Chattahoochee-Oconee National Forests, 1755 Cleveland Highway, Gainesville, GA 30501, (770) 297-3000

See also Georgia Bartram Trail Society and the NC Bartram Trail Society.

ment basins that were built and used as trout hatcheries in the 1930s by the CCC. Just past the abandoned hatchery site you'll see the trail-head for the Bartram Trail on your left, marked in Georgia with yellow diamonds.

You'll take a couple of switchbacks on the short climb to Warwoman Rd. Cross the road and then hike through thick rosebay rhododendrons (*Rhododendron maximum*) in a streamside acidic cove forest to cross a small bridge with a lovely vantage point for the cascading Becky Branch Falls at 0.3 mi. The canopy here is eastern hemlock (*Tsuga canadensis*); like all hemlocks in the Southeast, trees in this grove have been killed by an infestation of hemlock woolly adelgid (*Adelges tsugae*). The canopy is shared by tulip-tree (*Liriodendron tulipifera*).

The trail continues to ascend; just past Becky Branch, you'll reach a fork. Follow the yellow diamond-shaped blazes and take the left fork. However, if you are visiting in mid-spring, take a brief detour right and, immediately past the fork, examine the slope between the two trails, which has some large, deep red Vasey's trillium (*Trillium vaseyi*) as well as robust Solomon's seal (*Polygonatum biflorum*).

Returning to the yellow-blazed Bartram Trail, you'll find that the trail curves around to the left, climbing away from Becky Branch, and that the environment gets drier. The character of the forest changes from a second-growth cove forest that regenerated after logging (lots of second-growth tulip-tree and dying eastern hemlock in the canopy) to a young mixed pine/hardwood forest. You're walking along a south-facing slope, and if you look down at the bases of the trees here, you'll see the blackened evidence of fire. Understanding the human history of an area can often shed light on what you see on the ground. In this mixed pine/hardwood forest, the pines were killed either by southern pine beetles (*Dendroctonus frontalis*) or by fire. In the moister pockets, there are second-growth tulip-tree and eastern hemlock that regenerated following logging.

You might also notice the presence of Fraser magnolia, with whorls of improbably large leaves with distinctive ear-lobes at the leaf bases. Interestingly, Fraser magnolia was first described near this area by Bartram. Read his flowery description in the sidebar.

Three types of pine are common here: white pine (*Pinus strobus*), shortleaf pine (*P. echinata*), and Virginia pine (*P. virginiana*). If fire burned regularly enough to suppress the hardwood competition, this would be a fine example of a low-elevation pine forest. As it stands now, the rapidly growing hardwoods are on track to suppress the pines, shifting the balance toward oak dominance. The soil along the trail is dry, and the understory has mainly acid-loving species such as mountain laurel (*Kalmia latifolia*) and trailing arbutus (*Epigaea repens*), both in the blueberry family, Ericaceae. There is also a profusion of low-growing, maroon-flowered sweetshrub (*Calycanthus floridus*), which Bartram, in his *Travels*, described near this location as growing in "odoriferous groves." If you visit during the flowering season, be sure to catch a whiff of the overly sweet flowers and decide if you think they smell like cantaloupe, cherry lollipops, or something different entirely.

The trail crosses a dirt road about a mile from the trailhead, so look for the yellow diamond blazes to continue on the next section of trail across the road. You're now descending into a typical acidic cove forest, characterized by tall rosebay rhododendron and dead or dying eastern hemlock in the canopy. If you are here in the summer, you'll welcome the cooler microclimate as you cross a small bridge over the

Bartram's Mount Magnolia

The distinctive leaves of Fraser magnolia (*Magnolia fraseri*) were described in poetic detail by William Bartram as he traversed the southern Appalachians in 1775. He later called it *Magnolia auriculata* in his *Travels*, using the Latin root *auri-*, which means "ear." Bartram was referring to the distinctive "ear-lobes" at the bases of the large leaves. In botanical terms, we would refer to these leaves as auriculate. In Bartram's own words, "[The] fine flower sits in the center of a radius of very large leaves, which are . . . broad toward their extremities, terminating with an acuminated point, and backwards they attenuate and become very narrow towards their bases . . . with two long, narrow ears or lappets, one on each side . . . sitting very near each other, at the extremities of the floriferous branches, from whence they spread themselves after a regular order, like the spokes of a wheel, their margins touching or lightly lapping upon each other, form[ing] an expansive umbrella."

*Near this part of the trail, Bartram wrote, "This exalted peak I named mount Magnolia, from a new and beautiful species of that celebrated family of flowering trees, which here, at the cascades of Falling Creek, grows in a high degree of perfection" (*Travels *[1791]).*

creek. The humid air and tall vegetation might give you the sense of being in the forest primeval.

You'll begin climbing again out of the riparian area to enter an oak forest. Notice that the largest trees here, mostly white oak (*Quercus alba*) and white pine, are significantly larger than the trees you passed earlier, indicating that this section of forest is much older. The trail follows the left side of Martin Creek, then crosses on a small bridge

Towering white pine sentinels beckon the hiker through the older forest near the upper part of the hike, past the "stately columns of the superb forest trees," described by William Bartram in his Travels *(1791), and the roaring cascade of Martin Creek Falls, named Falling Creek by Bartram when he traversed this area in 1775.*

just before you reach a lovely backcountry campsite. Continue on the right bank until you cross back over the creek on a bridge to an overlook at the base of beautiful Martin Creek Falls.

Bartram gave the name Falling Creek to this tumbling cascade secluded deep in the shade of towering rhododendrons. Imagine yourself walking along with the great naturalist, who wrote, "I now entered upon the verge of the dark forest, charming solitude! as I advanced through the animating shades . . . between the stately columns of the superb forest trees, presented to view, rushing from rocky precipices under the shade of the pensile hills, the unparalleled cascade of Falling Creek, rolling and leaping off of the rocks: the waters uniting below, spread a broad, glittering sheet over a vast convex elevation of plain smooth rocks, and are immediately received by a spacious basin. . . . I here seated myself on the moss-clad rocks, under the shade of spreading trees and floriferous fragrant shrubs, in full view of the cascades."

Beyond Martin Creek Falls the Bartram Trail continues nearly 18 mi until it reaches the North Carolina border on NC 106, where it continues for nearly 75 mi in North Carolina. To return to the main Bartram Trail, retrace your steps and cross the creek, descending to the second bridge. Here, you would turn right to continue north on the Bartram Trail, past the campsite. However, we will turn left to return to Warwoman Dell. When you have caught your breath and replenished your energy reserves for the return hike, begin your descent. As you return, imagine yourself in Bartram's shoes, a lone explorer with a scientist's eye for botanical detail and a poet's gift of language.

6. STATION COVE FALLS

Why Are We Here?

Station Cove Falls is a perfect setting for the Goldilocks story of species diversity as it relates to forests in the mountains. You'll notice as you walk along the trail that some places are incredibly diverse – each time you look, you find more species. What makes some places "just right" for plant diversity? The rich cove forest at Station Cove Falls is a great place to think about how site conditions intersect with human disturbance and plant adaptations to result in areas with high diversity rather than dominance by a single or a few species. Station Cove is said to have the best wildflower displays in South Carolina, and a hiker visiting in late March or early April is sure to reap rich rewards in the form of colorful wildflower gardens.

The Hike

The trail begins behind the sign with the map. Descend through a pine forest that has extensive damage from southern pine beetle (*Dendroctonus frontalis*); follow the yellow blazes to walk parallel to Station Creek. You may notice a line of red- and blue-painted trees along the

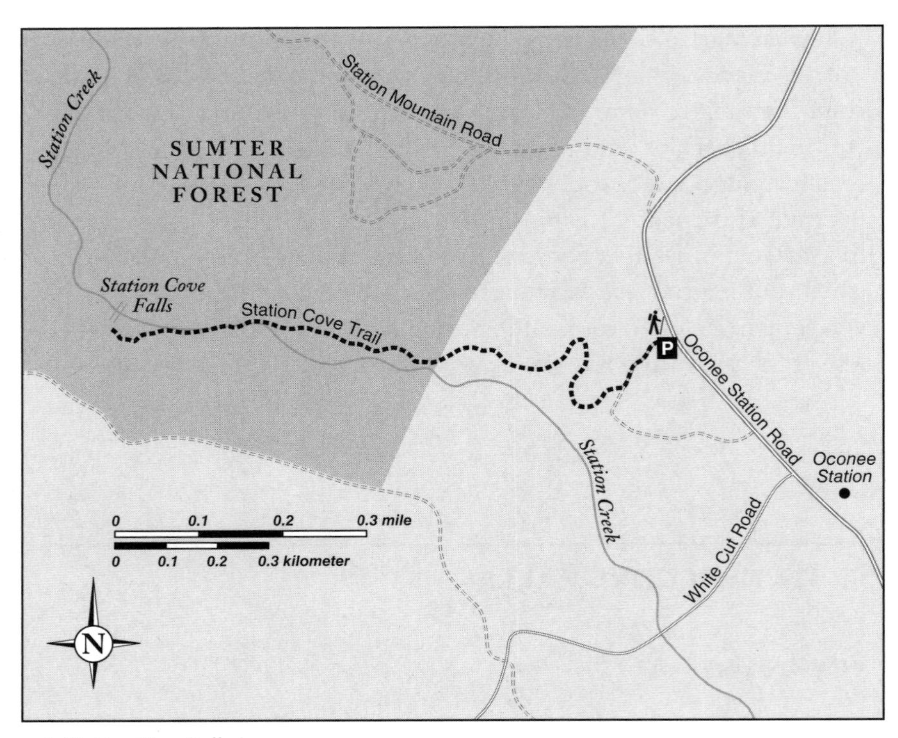

6. Station Cove Falls

way, which mark the boundary line between Sumter National Forest and the adjacent state-owned property. Below the trail you might also notice a swampy area that is a result of beavers damming up portions of Station Creek.

About a quarter of a mile along the trail, you'll bear left at a fork to reach Station Cove Falls; the trail that goes right is a connector trail that leads to Oconee State Park. This particular site has been designated as a Forest Service Botanical Area, which affords it the highest level of protection offered by the US Forest Service. You'll pass through a metal gate that limits the last 300 yards of the hike to foot traffic in order to protect the floral diversity of the Station Falls area.

As you follow the creek along this trail, notice that this is not a very old forest. Consider the largest trees here, some of which are recently dead pines. Pines tend to be the first colonizers of a pasture or agricultural field that has been abandoned, and these pines were

6. STATION COVE FALLS

Highlights: Incredible species diversity in a pretty cove forest, particularly spectacular in early spring, plus a lovely 60-ft waterfall at trail's end

Natural Communities: Cove forest (rich subtype)

Elevation: 1,140 ft

Distance: 1 mi round-trip

Difficulty: Easy walking on level trails

Directions: From the intersection of the Foothills Parkway (SC 11) and SC 28 at Walhalla, drive north on SC 11 for 6.3 mi. Turn left (north) on Oconee Station Rd. Drive 2.4 mi to the small parking area (4 cars only) and trailhead on your left. The parking area is about one-third of a mile past Oconee Station State Historic Site, where there are more facilities and additional parking. If you park there, a connector trail will lead you back to the Station Cove Falls parking area.

GPS: Lat. 34° 50′ 56″ N; Long. 83° 4′ 27.7″ W (trailhead)

Information: Sumter National Forest, Andrew Pickens Ranger District, 112 Andrew Pickens Circle, Mountain Rest, SC 29664, (864) 638-9568

at the end of their life cycle. Other larger trees you will encounter along this part of the trail are fast-growing tulip-trees (*Liriodendron tulipifera*), which grow quickly like pines but live much longer. This evidence of youth is further supported by an article published in 1857 in the *Keowee Courier* stating that from the porch of the Richards house (located at Oconee Station State Historic Site and built in 1805) you could see Station Cove Falls, a distance of about a mile. Additional documents show that this land was pasture at least through 1875.

This information is surprising because we usually associate wildflower diversity with old-growth forests. The absence of human disturbance, in the form of agriculture or logging, in old-growth forests allows disturbance-sensitive species, like delicate wildflowers, to thrive out of harm's way. This is, of course, an important part of the story—as you get farther back into the older forest, you'll see more species—but it's not the whole story.

Past the barrier that allows only foot traffic, notice as you walk along that some areas are dominated by a single species. On a recent visit in early April, we noticed a stretch of trail before the falls area that had an understory of nearly 100% mayapple (*Podophyllum pelta-*

Hyperfavorable pockets along the floodplain next to Station Creek are dominated by mayapple in early spring. Closer to the falls, look for a greater diversity of herbaceous plants.

tum). These plants are excellent competitors for sunlight in early spring, with their single or double umbrellalike leaves held parallel to the ground and their ability to grow in dense colonies. In this area, it's hard to spot bare ground. Mayapple is also somewhat resistant to disturbance, and it may have been able to recolonize more quickly than other, more sensitive species following a local disturbance. If mayapple were not here, there are doubtless many species that would thrive in this microsite. However, mayapple is well-established, and it thoroughly occupies its small piece of real estate, at least until the next disturbance.

Walk farther down the trail, and you will start hearing the roar of Station Cove Falls ahead. Here, look down and start counting the different species you see—how many can you count in a square yard?

This area of the trail, to the base of Station Cove Falls, has an incredible diversity of wildflowers. Why are there so many species? This is a much older forest (some estimate as old as 175 years) than the one you've walked through on your way here. The plant community has thus had more time to develop undisturbed. Another factor at play is the soil; many areas of the mountains and most of the Piedmont have acidic soils that are nutrient-poor. Here the forest grows over a mineral-rich parent material, and the pH is reported to be around 6.2. Less-acidic soil means that more nutrients, such as calcium and magnesium, are available for plant growth, and more species prefer neutral soils to acidic ones. The steep, rocky walls surrounding the falls also protect this cove from the harshest winter and summer temperatures. The bowl-like shape of the landscape retains soil and nutrients. Conditions are "just right" for many species—not too hot or too cold, not too dry or too moist, not too acidic or nutrient-poor. The herbaceous richness of this site raises some interesting questions. Is there an upper limit to the number of species that might coexist here? Could we artificially increase the soil fertility and continue to increase diversity? Ecologists have tried to answer these questions, which are explored further in this hike's sidebar.

When you've used your wildflower guide to identify as many species as you can, look around at the tree canopy at the base of the falls, an area that has much larger trees than the area you hiked through to get here. This forest is a fine example of a rich cove forest. While tulip-tree and oaks dominated the flats near the start of the trail, the trees near the falls are much larger, and there is no clearly dominant species. The tulip-tree, white ash (*Fraxinus americana*), basswood (*Tilia americana*), beech (*Fagus grandifolia*), and bitternut hickory (*Carya cordiformis*) are some of the largest examples in South Carolina, and the site conditions make them all robust competitors, with no single species dominating the canopy. One mountain cove species, yellow buckeye (*Aesculus flava*), is present in the subcanopy.

While these tree species have their own growth strategies and unique niches, they seem to be able to coexist in equal numbers around the lower falls. Thus we find that in rich cove forests like these, conditions also favor a large number of canopy species.

The stepped waterfall rises about 60 ft above you, creating a de-

Wildflower Diversity in Temperate Forests

Every southern Appalachian hiker will have noticed the great variation in richness or diversity (here we use these terms interchangeably) of herbaceous plant species across the landscape. Contrast the barren understories of many oak forests with the mind-boggling diversity of herbaceous species in a rich cove forest or northern hardwood forest. Such variation is one of the most striking differences among natural communities in the region, and its causes have led to much speculation among wildflower enthusiasts and plant ecologists alike.

The general consensus is that herbaceous species richness in temperate forests responds to differences in "favorableness" of the understory environment. Most forested understories are shaded, particularly during the growing season, so light levels on the forest floor are moderate and do not vary enough from one forest type to another to explain such striking differences in herbaceous diversity. What does vary considerably, however, is the ability of the forest floor to provide microsites for germination and establishment, plus abundant soil water and nutrients for plant growth. Stressful environments limit the number of species that can grow to those few hardy plants with specific adaptations to challenging conditions. Favorable environments make it possible for many species to thrive and coexist. Gardeners everywhere have drawn these same conclusions.

Next time you're out for a hike in the mountains, think about how the environment may affect herbaceous diversity. Dry, rocky sites have shallow soils that are drought-prone, while valley bottoms and floodplains have deeper soils with plenty of stored moisture. Plants will naturally thrive in the moister sites. Some forests have a thick litter comprised of acidic, decay-resistant, leathery leaves, like those of oaks or evergreen rhododendrons, while other forests have either bare soil at the surface or just a thin litter layer of rapidly decomposing leaves. Plants germinate and establish themselves readily in the latter sites, while a thick layer of acidic, decay-resistant mulch impedes germination and establishment. Soil fertility can also play an important role as well. Many rocks common in the southern Appalachians, such as granite and gneiss, produce soils that are acidic and nutrient-poor. However, some soil parent materials, such as limestone and rocks rich in the minerals amphibolite or olivine, produce soils that are dark, nutrient-rich, and of approximately neutral pH. Plants thrive in the richer, neutral-pH soils that develop on such rocks.

It should thus be no surprise that herbaceous diversity is higher in those "sweet spots" that combine adequate moisture, deep soils, and abundant nutrients. Because there is a continuous gradient of "favorableness" across the varied forest floor environments found in the mountains, it is not surprising that there is also a continuous gradient of herbaceous plant diversity. From this, you might conclude that herbaceous diversity should be highest in the best spots. However, ecological studies across a wide range of environments have shown repeatedly that the pattern is more complicated, and that "hyperdiverse" natural communities are rarely, if ever, found. Why? Up to a point, herbaceous diversity does increase with favorableness, but as the availabilities of water, soil, and nutrients continue to increase, herbaceous diversity begins to decline, almost as if there might be too much of a good thing. What actually seems to be happening, though, is that a few very effective competitors are able to monopolize the resources on extremely favorable sites and suppress weaker competitors.

While hyperfavorable sites are uncommon in the southern Appalachians, you may occasionally come across one. This will be a site with a deep, rich, dark, moist soil and relatively little surface litter in a place protected from the extremes of wind and the drying effects of sun, but not too deeply shaded. On such a site you may find yourself walking through a shoulder-high stand of rapidly growing, competitive herbaceous species such as touch-me-not (*Impatiens* spp.) or stinging nettle (*Laportea canadensis*). Few other herbaceous species will compete successfully with such fast-growing and site-occupying species.

The moral of this ecological story is that "just right" for herbaceous diversity is likely a case of "just enough," but not too much, of a good thing.

lightful blend of sight and sound as water splashes down to the pool at its base. Enjoy a rest at the base of Station Cove Falls and take in the beauty and variety of the tree canopy as well as that of the herbaceous understory. On your walk out, notice how the plant diversity on the forest floor decreases over this short, half-mile trail and try to imagine the view of the falls across the pasture that was once here before the Civil War.

As you enjoy Station Cove Falls, notice the concave, or bowl-shaped, landform, with the waterfall and Station Creek at the lowest point. Protected coves like this one offer favorable microhabitats for many species of wildflowers and trees.

7. FORK MOUNTAIN TRAIL

Why Are We Here?

For anyone who has spent time in the southern Appalachians, this hike offers a new way to see a familiar forest. Low- to mid-elevation oak forests occupy by far the greatest area in the southern Appalachians, found at all but the highest elevations and the most protected or exposed topographic positions. It's easy to overlook these widespread forests because they form the backdrop in the southern mountains; our attention naturally strays to the more exotic forest types that occur within this "oaky" background.

We'll focus in depth on these oak-dominated forests and the forces that shape them, as well as the subtle differences within these natural communities.

The Hike

Ellicott Rock Wilderness was set aside by Congress in 1975. More than 9,000 acres encompass the area where North Carolina (Nantahala National Forest), South Carolina (Sumter National Forest), and Georgia (Chattahoochee National Forest) meet. The historical convergence point for the three states is called Ellicott Rock, situated in the middle of the Chattooga Wild and Scenic River (the rock is not exactly at the official convergence point). We will not guide you all the way to the Chattooga River, but hiking there is an option if you're looking for an all-day excursion.

Several trail options exist at the trailhead parking area, so please stop at the kiosk and study the trail map. For the Fork Mountain Trail, you'll walk from the kiosk in the direction of the vault toilet, then continue in the same direction about 100 yards north along SC 107 from the parking area. Look for the trailhead marker on your left and enter the cool, damp tunnel of vegetation that begins this section of trail. The first part of the trail follows Dark Branch Creek, and the rush of water provides constant music as you walk along the gently rolling grade of the trail.

You are in what is known as the embayed area of the southern Blue

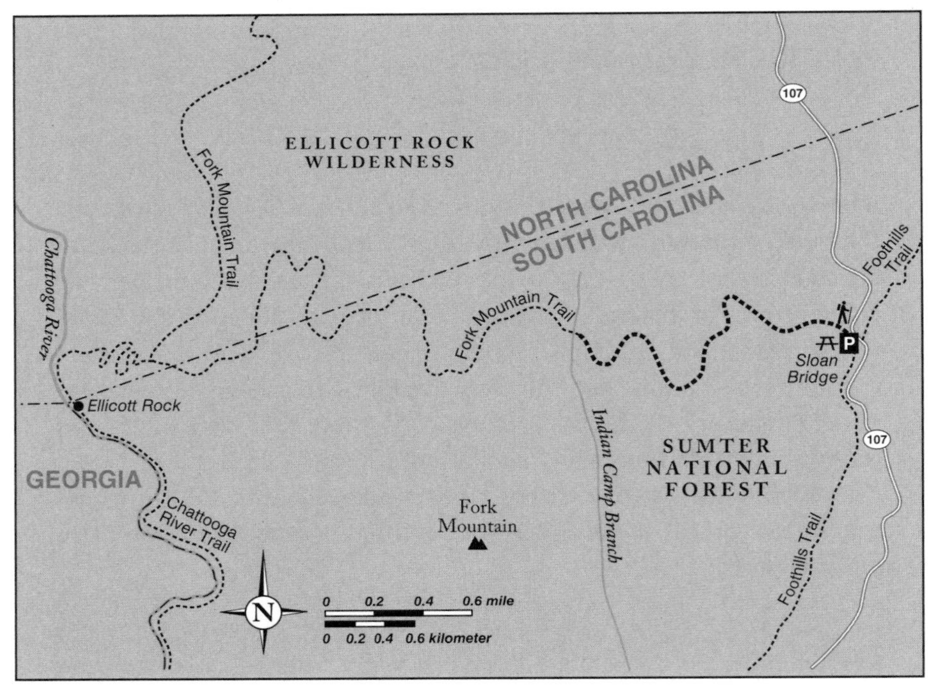

7. Fork Mountain Trail

7. FORK MOUNTAIN TRAIL

Highlights: Easy walking along rushing streams, past fern wonderlands, and through rhododendron and mountain laurel tunnels to a nice picnic spot along a secluded creek

Natural Communities: Oak forest, cove forest (acidic subtype)

Elevation: 2,800 ft at trailhead; no significant elevation gain/loss

Distance: 2 mi each way; 4 mi round-trip

Difficulty: Easy walking on well-marked foot trail along the river, sometimes wet and/or muddy but no significant stream crossings

Directions: From the US 64/NC 107 intersection in Cashiers, head south on NC 107. Cross the state line into South Carolina at 8.1 mi (road is now SC 107). Parking area is on the right at 9.0 mi (picnic tables, vault toilets).

GPS: Lat. 35° 0′ 12.3″ N; Long. 83° 3′ 15.1″ W (trailhead)

Lat. 35° 0′ 17.7″ N; Long. 83° 4′ 25.0″ W (Indian Camp Branch)

Information: Sumter National Forest, Andrew Pickens Ranger District, 112 Andrew Pickens Circle, Mountain Rest, SC 29664, (864) 638-9568

Ridge Mountains, carved into the southeast-facing escarpment. The embayment is an area of rugged topography, remote and undeveloped, with many rivers that drain to the southeast. You may learn more about this fascinating wilderness region by reading "The Southeastern Blue Ridge Escarpment" sidebar in hike 9 (Pinnacle Mountain). From the Chattooga River eastward, there are the Whitewater, Thompson, Horsepasture, Toxaway, and Eastatoe Rivers, all tumbling through steep-sided, rocky gorges. Of these rivers, only the Chattooga is navigable, but most sections of the Chattooga feature treacherous rapids and are therefore open only to skilled whitewater boaters.

After climbing some short stairs, you'll descend into a fern wonderland. The ground layer here is covered with two species, New York fern (*Thelyptris noveboracensis*) and hay-scented fern (*Dennstaedtia punctilobula*). Overhead, the canopy of this second-growth forest is surprisingly diverse, with many familiar trees, none of which is clearly dominant.

As you admire the ferns, look around at the forest. Note the distinct layering of vegetation, which ecologists refer to as strata. Can you identify layers of canopy, understory, shrubs, and herbs? Throughout this hike, pause to notice changes in these layers. Some places may not have all four strata, and you'll also notice shifts in species composition within each layer.

About a half mile from the trailhead, and not far past the fern wonderland, you'll descend and enter a tunnel of gnarled, tangled trunks. Look overhead and note that you are in an elfin forest dominated by nearly tree-sized rosebay rhododendron (*Rhododendron maximum*) and mountain laurel (*Kalmia latifolia*), both members of the heath or blueberry family, Ericaceae. There are a few true canopy trees that stretch over the top of this dense forest (you can only see the bases of their trunks here), but for the most part, shrubs form the dominant forest layer.

What conditions resulted in this unusual dominance by these heath species, where some of the individual trunks measure 6 to 8 inches in diameter, with root crowns easily 2 to 3 ft across? Unlike typical rhododendron thickets that reach 10 ft, this shrub canopy reaches 20 ft in places!

This area was logged at least twice before it was designated a wilderness. We also speculate that this area may have experienced selective cutting, where the large, valuable trees, such as American chestnut (*Castanea dentata*) and oaks (*Quercus* spp.), were removed, leaving a sparse canopy of a few trees with little commercial value, such as black gum (*Nyssa sylvatica*) and Fraser magnolia (*Magnolia fraseri*).

The thicket of tall heaths that now covers this area would inhibit the establishment of any new seedlings. Not only would these seedlings find little sunlight under the dense foliage, but decaying vegetation further acidifies the already acidic soils here, making an inhospitable rooting environment for many species. As a result, the rosebay rhododendron and mountain laurel were best able to benefit from the additional sunlight and nutrients that were freed up by the harvest of the canopy trees, and they are thus likely to dominate this site for many decades to come.

Just past this "heathy" tunnel, you'll reach a sign marking the boundary of Ellicott Rock Wilderness (0.6 mi). Be alert for large gaps in the canopy, many created by the death of eastern hemlock (*Tsuga canadensis*) caused by the hemlock woolly adelgid (*Adelges tsugae*). In addition, devastation by the southern pine beetle (*Dendroctonus frontalis*) of white and pitch pines (*Pinus strobus* and *P. rigida*) may be apparent. Peer upward into any gap, then look at the ground beneath and ask yourself which tree or shrub is best poised to capture this opening containing more sunlight, soil water, and nutrients. Think ahead and see if you can predict what the forest canopy will look like in 50 to 100 years. Will there be another elfin shrub forest, or will a tree canopy return here? There are a few places along the trail where you can see some very large trees, mainly oaks, which may have survived the initial logging. In these areas, you will have some insight into what this area may have looked like before most of the timber was cut.

As the trail undulates around dry ridges interspersed with ephemeral stream drainages, keep your eyes on the herb and shrub layers. You'll notice that rosebay rhododendron is replaced by mountain laurel as you move away from the creeks, and you'll notice differences in the herb layer as dominance shifts from mixed herbaceous species to mostly ferns. Other areas will have few herbs at all. Areas of sparse herbs on more open, south-facing slopes are often overtopped by a low, deciduous shrub called bear huckleberry (*Gaylussacia ursina*), also in the heath family.

Some of the ridges are drier, and you might detect subtle changes in

Ecologists call this transitional forest of yellow pine and mixed oaks, with bear huckleberry in the understory, pine-oak-heath. This is an oak forest in our classification, probably the most common forest type in the southern Appalachians.

canopy composition, where large chestnut oaks (*Quercus montana*) replace white pine. Degree of slope, aspect, and topography drive these subtle shifts. The driest areas have a convex shape, a southwestern exposure, and steeper slopes, while more mesic conditions are found in bowl-shaped areas with gentler slopes.

As you hike through the upland oak forests that cover most of this landscape, try to count how many different species of oaks you see. (Refer to the illustrations in the "Common Trees and Shrubs of the Southern Appalachians" section of this book.) If you tire of craning your neck to look up at the canopy, focus instead on the fallen leaves at your feet. Oak leaves resist decay, and intact specimens may be found in any season. A bouquet of oak leaves that you can identify will be a well-earned souvenir of your hike today! Look for white (*Quercus alba*) and chestnut oaks, with their rounded lobes, and red (*Q. rubra*), black (*Q. velutina*), and scarlet oaks (*Q. coccinea*), with their bristle-tipped lobes. For more information about our southern Appalachian oaks, please see the sidebar for this hike.

Past logging may have favored the establishment of a tall rhododendron understory at Ellicott Rock. Now, dying hemlocks create new openings, but the impenetrable shrub thicket below may mean that this forest of tall rhododendrons continues indefinitely.

You might ask as you hike through these lovely oak-dominated forests, why are the oaks so important here? This question has long puzzled both ecologists and commercial foresters. Oaks are among the most valuable timber species of the southern Appalachians; they are prized for their durable and beautiful wood and are suitable for paneling, furniture, and veneer. When oak-dominated forests are clear-cut for their timber, the trees that replace the oaks are often fast-growing species with lower commercial value, like tulip-tree (*Liriodendron tulipifera*), red maple (*Acer rubrum*), and white pine. Indeed, one of the biggest concerns among ecologists and foresters alike is that other species are replacing the once-common oaks in forests of the southern Appalachians.

As you continue your walk along the trail, note that there are few oak seedlings in the herb and shrub strata. Recruitment of oaks into these forests appears limited, and this raises interesting ecological questions: What conditions in the recent past favored the widespread growth of oak forests in the region, as encountered by the first European settlers and persisting well into the twentieth century? Why are other species now replacing oaks?

The answers to these questions are not clear, but Penn State University forest ecologist Marc Abrams thinks that fire suppression, which became effective in the early 20th century, may have played an important role in the current oak decline. Oaks seem to prosper best in forests that burn every 50 to 100 years, considered an intermediate fire frequency. If fires are too frequent, fire-adapted species, or even grasslands, predominate. When fires are too infrequent (which is the current situation), competitive thin-barked species such as red maple and Fraser magnolia may replace oaks.

After about 2 mi and several crossings of ephemeral creeks, you'll descend to a sweet little campsite situated by Indian Camp Branch, a small permanent stream. Pause here for lunch or a snack and, as you rest, take in the acidic cove forest around you. This is a common streamside community of the southern Appalachians, with an overstory of hemlock and white pine and a shrub layer of rosebay rhododendron, below which is another layer formed by a low, dense evergreen shrub, doghobble (*Leucothoe fontanesiana*). Around this campsite, notice that the areas of doghobble are always close to the water.

We are currently witnessing a dramatic change in this streamside

Southern Appalachian Oaks

Learning about oaks should be high on anyone's to-do list if they wish to become more knowledgeable about southern Appalachian forests. A good starting point is learning to distinguish the white and red subgenera in the large genus (*Quercus*) that contains the species of oaks. We separate white, post, and chestnut oaks from all the others for a reason: these three species belong to the "white" subgenus of oaks, while the others are members of the "red" subgenus. For the finer points of tree identification, we recommend that you begin with the illustrated guide in this book and then consult one of the many other excellent guides available. The oaks can be challenging to learn, and within the subgenera, the species are not shy about producing the occasional hybrid; so don't feel bad if you cannot identify every leaf you find.

Even within a given individual, oaks (and other plants, for that matter) can vary considerably in leaf form. Leaves of seedlings and saplings, and those produced in the shade on larger specimens, are typically larger and

community with the catastrophic loss of eastern hemlock. The hemlock woolly adelgid, an exotic pest introduced from Japan to the Pacific Northwest, has slowly spread to the East Coast and is decimating these forests. Since hemlock dominates this forest canopy, ecologists wonder what species will take its place and how these communities will change as the hemlocks fall. Already, small streams are experiencing a temperature increase as additional sunlight reaches them. Mountain stream dwellers, such as the native brown trout (*Salmo trutta*), are threatened by these changes in their environment. You can learn more about this situation in the "Hemlock Woolly Adelgid" sidebar in hike 18 (Ramsey Cascades).

Just past the creek, you can see a dead and dying canopy of even-aged white pine (killed by the native southern pine beetle) interspersed with some dead hemlock. How do you know the white pine is even-aged? If you look around the forest here, you'll see that most of these trees are about the same size. On a recent visit, we measured several white pines in the range of 16 to 24 inches in diameter at breast height (dbh, or height at 4.5 ft, the standard place of measure-

have less-pronounced lobing than the "sun" leaves produced higher in the canopies of larger individuals. Ecologically, this makes sense: in the shade, larger leaves are needed to capture as much sunlight as possible. At the top of the canopy in full sunlight, the bigger problem for leaves is drying out, and, accordingly, leaves are smaller and thicker to reduce water loss through evapotranspiration. They are also more deeply lobed to assist in convectional cooling.

Once you begin to feel comfortable with identifying oaks, try to associate the various species with different habitats. For example, scarlet and black oaks favor the drier, steep, and south-facing slopes; chestnut oak often occupies open slopes of intermediate dryness, while red and white oaks will be found on slightly moister sites in the landscape. The latter two species will also become abundant at higher elevations, as you shall see on several of our hikes. However, please be aware that most of our oaks are generalists, meaning they can tolerate a broad range of environmental conditions.

ment for foresters). White pines tend to establish in open areas, so we can speculate that this area was logged and then grew up in a second-growth stand of white pine after it was clear-cut.

Unlike white pine, eastern hemlock can tolerate shade, so the hemlock could have established itself under the white pine, though some individual trees are just as large as the white pines. Now, both species are dead or dying in this stand, raising the question of what the canopy will look like in the future. As you look around, what species do you see capturing the canopy gaps? We observed species such as Fraser magnolia, red maple, and even American holly (*Ilex opaca*) in some of the reforested gaps.

Past the campsite at Indian Camp Branch, the Chattooga River is another 4.5 mi. If you plan to continue to the river, keep an eye on the weather and make sure you have plenty of water, snacks, and available time and daylight for your return. However, this is the terminus of our guided hike in Ellicott Rock Wilderness, so from here you'll turn around and return along the same trail to the parking area. As you retrace your steps, consider how the forces of past logging, fire suppression, and current pests (both native and introduced) have shaped this forest, and think about what this forest might look like 100 years into the future.

8. OCONEE BELLS NATURE TRAIL

Why Are We Here?

Devils Fork State Park offers two loop hikes, Bear Cove Trail (2.5 mi) and Oconee Bells Nature Trail (1.0 mi). Both trails feature easy access to large streamside populations of a low-growing evergreen plant, *Shortia galacifolia*, also known as Oconee bells. This is one of the rarest plants of the southern Appalachian Mountains, and it is an excellent example of a narrow endemic, a species found only in a very limited area within a particular geographic region. If you are able to visit in early March to mid-April, you will have the added treat of seeing this lovely little plant topped with white and pale pink bell-like flowers. The Oconee Bells Nature Trail also provides an opportunity to view a typical oak forest. This forest is similar in its structure and composition to stands throughout the Piedmont province to the south and east.

The Hike

The Oconee Bells Nature Trail starts at a kiosk at the southeastern corner of the large parking area away from the lake. After descending some steps and traversing the graveled path below, you will arrive at the junction of the loop hike. Take the right fork, which descends gently through a forest of oaks and pines (low-elevation pine forest) to a stream, where you will be following white trail blazes. On this section of the trail, white pine (*Pinus strobus*) is prevalent, along with white oak (*Quercus alba*), with its round-lobed leaves. Closer to the stream, you'll see rosebay rhododendron (*Rhododendron maximum*) dominant. Take a look at the ground – there is very little in the understory along this initial stretch of trail.

However, as you follow the stream, look for the round, glossy, dark

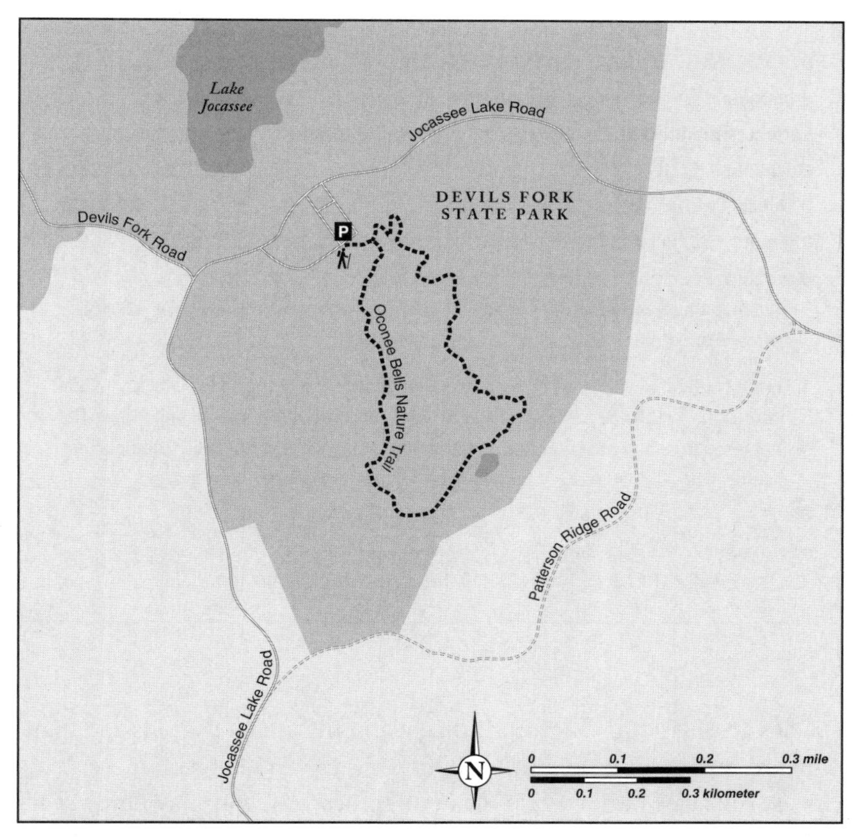

8. Oconee Bells Nature Trail

green leaves of Oconee bells forming dense mats on both banks. If you are hiking in March or early April, you will see this attractive plant in full bloom, with individual nodding flowers borne on 4- to 6-inch stems above the foliage. The distinctive leaves of Oconee bells very much resemble those of the related and widespread galax (*Galax urceolata*), hence the Oconee bells' specific epithet, *galacifolia*, meaning "galax-leaved."

Oconee bells' native range is restricted to the drainages of just a few rivers cascading down to the Piedmont from the southern Blue Ridge Mountains, in Transylvania and Jackson Counties, NC; in Oconee and Pickens Counties, SC; and in Rabun County, GA. An outlying population on tributaries of the Catawba River is found in McDowell County, NC, near the town of Marion. Oconee bells' closest relatives are other

8. OCONEE BELLS NATURE TRAIL

Highlights: Easy access to large population of Oconee bells (*Shortia galacifolia*)

Natural Communities: Low-elevation pine forest, oak forest

Elevation: 1,200 ft

Distance: 1 mi on Oconee Bells Trail (loop)

Difficulty: Easy walking on well-marked foot trail

Directions: From I-85 at the NC/SC line, travel south on I-85 to Hwy 11 exit at Gaffney. Proceed on Hwy 11 for approximately 90 mi. Turn right on Jocassee Lake Rd. and travel 3 mi to the park.

From I-26, take exit 5 onto Hwy 11 to Campobello. Travel for approximately 60 mi. Turn right on Lake Jocassee Rd. and travel 3 mi to the park. The Oconee Bells Trail begins in the southeast corner of the parking lot adjacent to the visitor center (restrooms, store, picnic shelters, camping, rental cabins/villas, boat ramps).

GPS: Lat. 34° 57′ 5.9″ N; Long. 82° 56′ 47.1″ W (trailhead)

Information: Devils Fork State Park, 161 Holcombe Circle, Salem, SC 29676, (864) 944-2639, (866) 345-7275 (toll-free), devilsfork@scprt.com

species in the genus *Shortia* found in mountainous regions of Japan, Taiwan, and mainland China. If finding an endemic plant with its closest relatives half a world away intrigues you, you may enjoy reading the "Eastern North American–Eastern Asian Connections" sidebar in hike 14 (Joyce Kilmer Memorial Forest). The dark green evergreen foliage and attractive white to pinkish flowers of Oconee bells have made this a popular plant with gardeners, who have introduced it to many areas well beyond its native range.

You will notice how the Oconee bells are restricted to the vicinity of the stream channel. The combination of relatively abundant and reliable moisture along with higher light levels seems to favor the species in this microhabitat. You will also find many seedlings along the banks where flooding has exposed fresh soil. It seems likely that the drier, shadier habitat of the adjacent forest, with its soil covered in a thick layer of decaying oak leaves, inhibits both reproduction and growth of the Oconee bells.

After enjoying the display of Oconee bells, continue your hike. The trail mostly hugs the stream, occasionally rising into low woods. Af-

The Historical Mystery of Oconee Bells

Delicate and shy, Oconee bells eluded botanists for nearly 100 years before being rediscovered along low-elevation streams in the southern Appalachians, like those at Devils Fork State Park, where it grows prolifically.

The rare and narrowly distributed Oconee bell (*Shortia galacifolia*) was discovered by someone who didn't name it and named by someone who couldn't find it after someone who never saw it. Botanist and explorer André Michaux discovered and collected the plant in 1788, and the specimen remained in a herbarium in Paris until Harvard botanist Asa Gray came across it in 1839. Gray named the plant to honor American botanist Charles W. Short, but Gray had never seen it in the wild, having only Michaux's vague location information on the specimen (*les haute montagnes de Carolinie*, or the high mountains of Carolina) as a guide to its location. Gray made it his quest to find the plant, making several unsuccessful forays to the southern mountains in pursuit of the "lost plant of the southern Appalachians."

The eventual discovery of a natural population of Oconee bells was made possible by the alertness of a 17-year-old amateur botanist and plant collector named George Hyams, who found a population of the plant near Marion, NC, in 1877. Asa Gray eventually saw Oconee bells in the wild in 1879, although he never visited the site of Michaux's original collection. Indeed, Michaux described this site in greater detail in his field journal and clarified that the plant was found in low elevations on the banks of a stream, and therefore unlikely to be in the high mountains where Gray had been searching.

A sad footnote to this story is that the original collection site (in Oconee County, SC, source of Michaux's type specimen) was lost when Lake Jocassee was created, flooding this site and 60% of the global population of Oconee bells.

ter the trail crosses the stream and begins looping back toward the parking area, you will walk through an area that is predominantly an oak forest. You will pass a small pond to the right with evidence of recent beaver (*Castor canadensis*) activity—look for the irregular, pointed remnants of gnawed small trees near the water. When you get a clear view of the pond, see if you can spot a mound of sticks and branches, the lodge constructed by these furry engineers. Beavers occupy wetland areas and are a keystone species because of their instinctual behavior to halt flowing water by building dams, in turn flooding and creating more wetland habitat.

After leaving the pond area, you will climb gradually through a young oak forest on your return to the parking area. Look for the fuzzy, pointed lobes of the leaves of black oak (*Quercus velutina*), which dominates here, along with various other oaks and hickories. If you are familiar with oak forests of the Piedmont, you will feel right at home in this forest in the foothills of the Blue Ridge Mountains.

9. PINNACLE MOUNTAIN

Why Are We Here?

If you are willing to lace up your sturdy boots, pack a lunch, and take on a challenge, the Table Rock and Pinnacle Mountain Trails will reward your hard climb with sweeping views of the South Carolina Piedmont. Along your journey, you may be surprised to find giant boulders and remarkably varied forests. You will summit Pinnacle Mountain, sometimes called the REAL highest point of South Carolina (Sassafras Mountain holds the honor of the highest point, but since it straddles North and South Carolina, it is not always claimed by SC).

Indeed, the story here is one of two mountains, and you will see their geologic differences reflected in the forests that cloak their summits. Table Rock is composed of granite rock at its summit, but the summit of Pinnacle Mountain is underlain with amphibolite, which produces a less acidic soil with more available nutrients. We don't

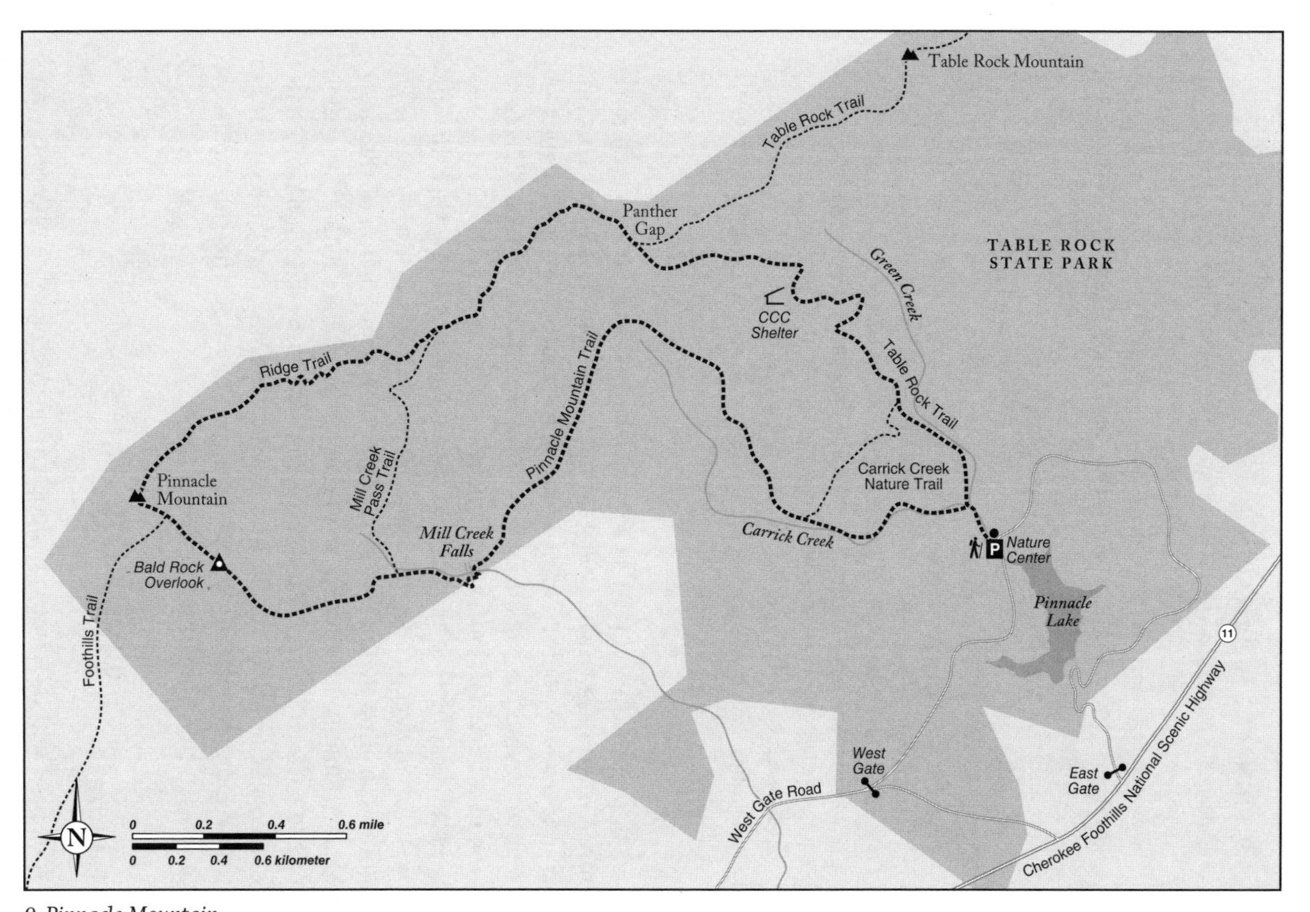

9. Pinnacle Mountain

9. PINNACLE MOUNTAIN

Highlights: A challenging hike that spans two mountains, featuring huge boulders and panoramic views south into the Piedmont

Natural Communities: Cove forest (acidic subtype), oak forest

Elevation: 1,170 ft (trailhead); 2,552 ft (Panther Gap); 3,425 ft (Pinnacle Mountain summit)

Distance: 8.2 mi, connecting several trails to form a loop

Difficulty: Strenuous; some steep climbs up Table Rock Mountain and equally steep descents down Pinnacle Mountain

Directions: From the Foothills Parkway (SC 11), take the East Gate entrance to the park, turning north (if you turn south, you will arrive at the visitor center, where you can obtain a park map). Follow the road past the lodge and follow signs toward the lake. Park in the large parking area next to the lake. Cross the road to the interpretive center and restrooms. The trailhead is just behind this building.

GPS: Lat. 35° 1′ 9″ N; Long. 82° 41′ 31.7″ W (trailhead)

Information: Table Rock State Park, 158 E. Ellison Ln., Pickens, SC 29671, (864) 878-9813

need to perform soil tests to see this—the evidence is reflected in the forest, if you know what to look for. Our ecological goal for this hike is to introduce you to some of the effects that varied geologic parent materials, and the soils that form from them, can have on the vegetation of otherwise similar landscapes.

From reading this brief introduction, you now know that this hike is set in a rugged mountainous terrain that lies just adjacent to the South Carolina Piedmont. You'll be exploring a portion of the southeastern Blue Ridge Escarpment, one of the richest biological regions of the eastern United States, a region shared by the adjoining states of Georgia, South Carolina, and North Carolina. If you are curious to know more about the escarpment, you may find the sidebar associated with this hike of particular interest.

The Hike

Your hike will take you on a loop comprised of three different trails. You will start by following the red-blazed Table Rock Mountain Trail. At Panther Gap, you'll turn left and follow the orange blazes along the Ridge Trail to reach the summit of Pinnacle Mountain. Then you'll pick up the yellow-blazed Pinnacle Mountain Trail for your descent back to the trailhead.

Before starting your hike, please stop at the kiosk and register your planned route. Begin your journey on the popular Table Mountain Trail behind the Carrick Creek Interpretive Center, walking up the paved pathway and crossing a bridge over the confluence of Green Creek and Carrick Creek. At the trailhead and in this very first section of trail, you'll notice beautiful stonework along the creek. This is the work of the Civilian Conservation Corps, which constructed many park facilities during the Great Depression. The first section of trail is blazed with green, red, and yellow, and you'll soon reach a Y after crossing over Green Creek, where you will follow the red and green blazes to your right on the Table Rock Trail.

You'll notice subtle differences in the forest here, depending on how close you are to Green Creek. Next to the creek, you'll be in an acidic cove forest dominated by Eastern hemlock (*Tsuga canadensis*), but farther away you'll be in an oak forest with a more diverse mix of trees dominated by shortleaf pine (*Pinus echinata*), some white pine (*P. strobus*), but also white oak (*Quercus alba*), black gum (*Nyssa sylvatica*), sassafras (*Sassafras albidum*), and sourwood (*Oxydendrum arboreum*). The presence of pines here might indicate that this part of the forest is younger and has been disturbed by fire or logging. Supporting this premise is the fact that the trees here are not especially large.

In both forest types (acidic cove forest and oak forest), you'll notice that acidic soils preclude the growth of most wildflowers. Instead, the tree and shrub layers dominate. Near the stream you'll see more rosebay rhododendron (*Rhododendron maximum*), and farther from the stream you'll see more mountain laurel (*Kalmia latifolia*) and a small-leaved, evergreen ground cover, partridgeberry (*Mitchella repens*).

After another crossing of Green Creek less than a half mile from the trailhead, you'll climb through a steep-sided ravine, with the creek to your left. At the top of this section, a lovely waterfall provides background music, inviting you to take a break. Here you might notice that the character of the forest changes a bit. Though the canopy trees are still oaks and hickories, with an occasional tulip-tree (*Liriodendron tulipifera*), the evergreen shrubs have disappeared. Just past the half-

The Southeastern Blue Ridge Escarpment

Throughout most of their length, the Blue Ridge Mountains present a formidable, steep, and predominantly east-facing escarpment, which forms the abrupt boundary between the gentle, rolling topography of the foothills and Piedmont to the east and the Appalachian Mountains to the west. Toward its southern terminus, however, the Blue Ridge Escarpment becomes increasingly southeast- to south-facing, and more diffuse. Rather than a single massive wall, it breaks into a choppy sea of peaks, ridges, steep slopes, and deep valleys known as the southeastern embayment.

The climate of the embayment is relatively cool, of course, but milder than that in the heart of the southern Appalachians. Moist subtropical air surging up from the Gulf of Mexico collides with the steep topography here, creating some of the highest precipitation found anywhere in North America east of the Cascade Mountains of the Pacific Coast. Numerous rivers cascade through steep gorges, racing through deep chasms and tumbling over steep cliffs.

The southeastern embayment region (or Blue Ridge Escarpment) comprises the upstate regions of Georgia and South Carolina and adjacent parts of North Carolina. Because of its remote and rugged nature, the region has been spared the development that has overtaken so much of the southern Appalachian region in the past century.

From a biological perspective, the southeastern Blue Ridge Escarpment is a "hot spot" of biological diversity in eastern North America. For South

mile marker, you'll reach the intersection where the Carrick Creek Trail heads off to the left; you'll continue on the red-blazed Table Rock Trail up and to your right.

During the growing season, you might notice a shrub that looks quite like your garden-variety hydrangea, with large, opposite, rounded leaves. Turn one of these leaves over to see its bright white surface. This is a wild, native hydrangea, *Hydrangea radiata*, which is only found on the Blue Ridge Escarpment. During the growing season, you may see scattered wildflowers and no sign of acid-loving rhododendron or mountain laurel. Their absence might be a sign that the soil here is different from that on the rest of the Table Rock Trail.

Carolina alone, the Nature Conservancy, in its "South Carolina Blue Ridge Escarpment" (2013), reports that, although comprising less than 2 percent of the state's area, this region provides "habitat for 7 plants and one animal on the federal list of endangered and threatened species . . . more than 300 rare plants and animals that are tracked by the SC Heritage Program, including more than 150 plants, 24 mammals, 5 amphibians, 5 reptiles, 4 birds and a few mosses, liverworts, lichens, mussels and caddisflies . . . [and] more than 40 percent of the plants on the South Carolina rare plants list."

Amphibians, such as this young red-spotted newt, highlight the importance of conserving biological diversity in areas along the Blue Ridge Escarpment, such as Table Rock State Park.

Outside this ravine, the understory reverts to mainly mountain laurel with some rhododendron for the rest of your climb to Panther Gap.

Just past the 1-mi mark, the trail switchbacks and climbs steeply, and you will once again be surrounded by giant granite boulders. The forest on this mainly south-facing, low-elevation slope has to cope with drought and hot summer temperatures. Accordingly, the trees that grow here are adapted to the dry, acidic, and thin soils. Still in an oak forest, you will see as you climb quite a bit of scarlet oak (*Quercus coccinea*) and chestnut oak (*Q. montana*), interspersed with pockets of Virginia pine (*Pinus virginiana*).

You'll reach a stone shelter where you can take a well-earned break just past the 1.5-mi marker. Pause and enjoy the views below; there's an even more open view just up from this shelter where you can easily see Pinnacle Lake and, beyond that, Lake Ooleny (where the main park office is) stretching out far below you. It's hard to imagine that you have climbed all that way! The overlook is at 2,400 ft, so you have gained roughly 1,200 ft in elevation already. Before you continue your climb to Panther Gap, scan the sky to make sure you're not in for any afternoon thunderstorms.

You will reach Panther Gap (elevation 2,552 ft), the lowest point between Table Rock and Pinnacle Mountain, just past the 2-mi mark. Most hikers follow the trail to the right to continue to the summit of Table Rock, climbing another 500 ft, for more spectacular views and more of the same forest type. However, we will take the road less traveled, following the orange blazes to the left along the 1.9-mi Ridge Trail, which leads to the summit of Pinnacle Mountain.

The trail follows the ridgeline, and two things might catch your eye immediately: first, you have additional plants in the herb layer (on a recent visit in early April, bloodroot [*Sanguinaria canadensis*] abounded in white carpets, with single, Matisse-inspired leaves), and second, there are more large trees. Many of these large trees are northern red oak (*Quercus rubra*), with some large white and scarlet oaks intermixed. You'll also notice more tulip-tree in the canopy here than you saw earlier.

This oak forest with abundant herbs indicates that rich soil underlies this section of trail. If you observe the bare mineral soil on some sections of trail, you might notice that the color of the soil is a rich brown, rather than gray or reddish. Soil color changes, however, as you move out of Panther Gap, so try to see if you can notice this transition in the color of the rocks and the soil beneath your feet. For the most part, the Ridge Trail follows the southern side of the ridgeline, which faces southeast, as it leads to the Pinnacle Mountain summit.

As you traverse the ridgeline, notice as you ascend and descend that you'll go through patches of drier forest with tunnels of mountain laurel and rhododendron, indicating acid soils, interspersed with areas of large oaks and an herbaceous, rather than shrubby, understory. All of this is considered oak forest, but depending on the soil type, soil depth, slope, and aspect, its character can change rather dramatically. Ecologists might further parse these differences in forest cover into more specific communities. What we want you to see is how much variation there is within the concept of oak forest.

One mile past Panther Gap along the Ridge Trail you'll reach Mill Creek Pass, where a blue-blazed trail cuts left (south) to join Pinnacle Mountain Trail below the summit. If it is getting late in the day, this shortcut will save you about a mile. You will turn left when you reach the Pinnacle Mountain Trail to head back toward the parking lot. If you have the time and the weather is favorable, simply continue on the Ridge Trail.

From Mill Creek Pass, you'll have just under a mile of steady climbing over two more peaks before your final ascent to the summit of Pinnacle Mountain at 3,425 ft. Again, the top of this mountain is underlain with amphibolite, so you may see some wildflowers in the understory at the peak and on your initial descent.

There is no view from this forested summit, so linger only briefly before you begin the steep descent to your left down the yellow-blazed Pinnacle Mountain Trail. You'll lose nearly 400 ft of elevation in two-tenths of a mile before you reach the intersection with the Foothills Trail. Take your time and notice the profusion of plants in the herbaceous layer here, with the underlying brown soil indicating the amphibolite parent material and richer soil.

The Foothills Trail branches off to your right, marked with white blazes, but you will continue your steep descent to the Bald Rock Overlook (elevation 2,800 ft) for a rest and a panoramic view to the southeast into the South Carolina Piedmont. If you continue down the trail to the edge of the overlook, you can see Table Rock, and you will have excellent views of the lakes below and to your left. Tread carefully to avoid crushing the lichens and small plants that cling to the surface of the exposed rock.

Below, the rock reverts mainly to granite, so you will gain shrubs at the expense of herbs for much of the rest of your hike. The descent to the intersection with the Mill Creek Pass Trail (blue blazes) is a bit more gentle. At the first intersection with that trail, you will see a 0.4-mi spur trail leading to Mill Creek Falls, a lovely diversion if you have time. Otherwise, continue to follow the yellow blazes, and you'll come to a huge rock outcrop and overhang, just before another waterfall adjacent to the trail. On a hot day, the cool spray is a welcome treat.

As you descend, more gradually now, along the Pinnacle Mountain

Descending from the summit of Pinnacle Mountain, you'll soon reach Bald Rock Overlook and its lovely views. As the trail turns left and continues past the first overlook area, you may spot the Table Rock summit looming in front of you on a clear day.

Trail, you'll observe the renewed presence of large boulders along the trail. Notice, too, that the forest next to the creek is once again an acidic cove forest, made apparent by the thickets of rhododendron and renewed presence of eastern hemlock.

When the trail intersects with the Carrick Creek Trail, you will walk alongside Carrick Creek, a popular spot for families enjoying the water, then hopscotch across large stones to span the creek twice before you reach the intersection with the Table Rock Trail to complete your loop hike. At this intersection, turn right to retrace your steps to the small building and the parking area. Don't forget to sign out at the kiosk so park rangers know that you returned safely from your adventure! Once you reach the lake, turn around to gaze up to Panther Gap, the low point between Table Rock and Pinnacle Mountain, and then the summit of Pinnacle Mountain itself, to mentally retrace your steps across these two very different mountains.

10. RAINBOW FALLS

Why Are We Here?

Jones Gap State Park has a variety of landform features, and here you can get a feel for the ever-changing landscape of the embayed area of the Blue Ridge Escarpment. You can experience the rugged terrain of rockslides and waterfalls, with opportunities for hikes ranging from relatively easy to extremely strenuous.

The hike to Rainbow Falls includes examples of how dynamic forces—earth, water, and fire—change the trajectory of forest development. We will consider the role of these primordial forces in shaping vegetation. Waterfalls, wildlife viewing, and diverse forests will all be part of your experience here.

The Hike

Jones Gap State Park (3,346 acres) is a gateway to the 11,000-acre Mountain Bridge Wilderness that includes Jones Gap and Caesar's Head State Parks. Starting at the Jones Gap Learning Center, cross the Middle Saluda River on a broad wooden bridge to reach the trailhead of the Jones Gap Trail, recently designated as part of the Palmetto Trail, which traverses South Carolina. Be sure to enter the requested information in the logbook here; this is required of all hikers entering the Mountain Bridge Wilderness. The trail follows the Middle Saluda River along the left bank as you hike upstream. The first part of the hike is easy going, following the bed of an old toll road through Jones Gap and marked with blue blazes.

You'll pass the intersection of the Rim of the Gap Trail in a quarter mile on your left; continue to follow the Jones Gap Trail, taking time to pause at some of the openings along the riverbank to admire the tumbling Middle Saluda River. The forest, too, is interesting, comprised of low-elevation species that you would expect to see, such as American sycamore (*Platanus occidentalis*), with surprising occurrences of species you would typically find only at higher elevations, such as basswood (*Tilia americana*), with its large, heart-shaped leaves, and Fra-

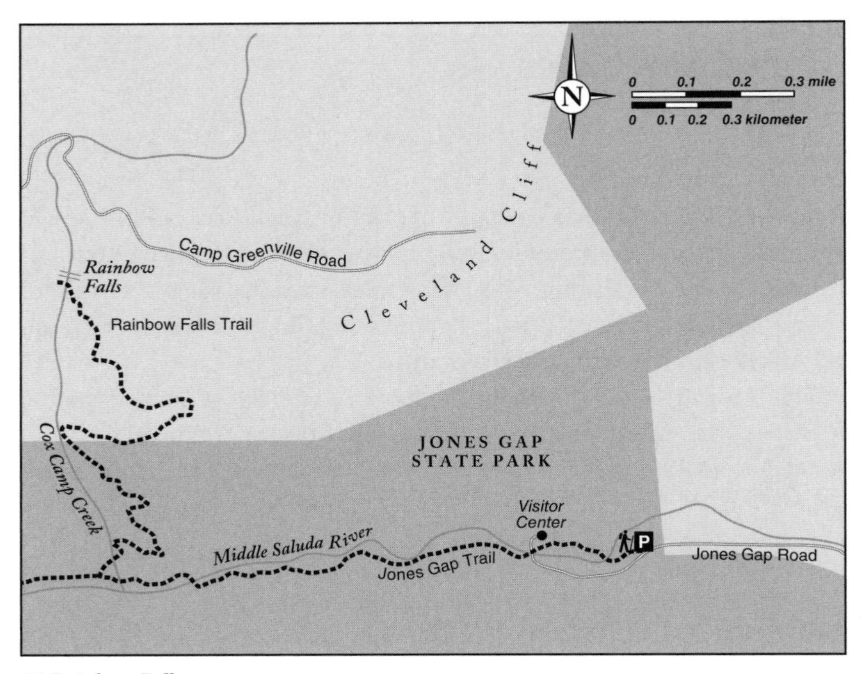

10. Rainbow Falls

ser magnolia (*Magnolia fraseri*). It's hard to choose a single dominant species; this forest contains a diverse mix.

Along the way you will pass hike-in campsites along the river; these can be reserved in advance through the park office. Just past campsite #9, slightly over a half mile past the trailhead, the trail climbs and then widens into a flat area covered with tumbled rocks. You will see an intermittent small stream coming through the trees to your left; during times of heavy rainfall, this stream floods the trail, and you'll have to cross it using the rocks on the trail as stepping-stones.

What you are seeing here is a massive rockslide that occurred in the late 1970s. This landslide shook the earth enough to register on seismographs in Columbia, SC. As you cross this area, consider the consequences of such an event for the forest community. While evidence of the immediate mortality suffered by trees in the rockslide's path is waning after a few decades, you may be able to discern how other trees took advantage of this disturbance.

Pause to note the principal beneficiaries of this catastrophic event: tulip-tree (*Liriodendron tulipifera*), American sycamore, and sweet birch

10. RAINBOW FALLS

Highlights: A moderate hike through cove forest along the beautiful Middle Saluda River, a pristine mountain stream that tumbles over boulders, offering fine opportunities for the trout fisherman. Waterfalls and landslides remind you that you are in the dynamic landscape of the southeastern Blue Ridge Escarpment, one of the largest wilderness areas in eastern North America.

Natural Communities: Cove forest (acidic subtype)

Elevation: 1,400 ft (Jones Gap Learning Center); 2,800 ft (Rainbow Falls)

Distance: 4.9 mi round-trip

Difficulty: Moderate for the first mile, with a steady and sometimes steep climb on the Rainbow Falls Trail

Directions: From Greenville, SC, take Hwy 276 through Travelers Rest, Marietta, and Cleveland. Approximately 2 mi north of Cleveland, turn right at the park sign onto River Falls Rd. River Falls Rd. changes to Jones Gap Rd. Continue to the parking area loop, which fills quickly on weekends. After parking, walk 0.2 mi across the Middle Saluda River on a wooden bridge to reach the park visitor center and the trailhead (self-service fee kiosk, accessible parking, picnic area).

GPS: Lat. 35° 7′ 31.8″ N; Long. 82° 34′ 9.9″ W (parking)

Information: Jones Gap State Park, 303 Jones Gap Rd., Marietta, SC 29661, (864) 836-3647

(*Betula lenta*). Each species has a different growth strategy that has contributed to its success in this disturbed area. Tulip-tree grows most rapidly; its airborne seeds germinated quickly in the open areas with abundant sunlight after the landslide. You might notice that American sycamore, the most flood-tolerant of the three beneficiary tree species, grows most abundantly close to the Middle Saluda. Finally, birches can establish themselves from seed on a variety of substrates that offer some moisture and freedom from competition. These substrates can include rotting logs, mossy boulders, and in this case, bare soil exposed by the rockslide.

Note the many pole-sized trees racing skyward for a share of the abundant sunlight in the gap. Not all will survive, of course, but those that do will form part of an even-aged stand that will bear witness to this event for the next century or more—or until the next rockslide!

Rockslides are not uncommon in the southeastern Blue Ridge Escarpment, a region also well-known for its abundance of waterfalls. You have just seen one of two massive rockslides that can be viewed

Tumbled rocks and young birch, tulip-tree, and sycamore trees mark the site of a massive rockslide that occurred in the 1970s. Notice that the trees are all about the same diameter, a clue that a large-scale disturbance regenerated this young forest.

along this hike (the other, which is harder to see, is across the Middle Saluda; it occurred in July 2006 after a 16-inch rainfall), and your final destination will be a beautiful waterfall, Rainbow Falls. The sidebar for this hike discusses the proliferation and causes of features like waterfalls and rockslides in this region.

Just a tenth of a mile past the old landslide, you'll come to an intersection where the blue-blazed Jones Gap Trail continues to the left, while the Rainbow Falls Trail continues straight ahead, with red blazes. Here you'll leave the Jones Gap Trail to follow the Rainbow Falls Trail, which was established in 2008. A short distance past the intersection, you'll take a wooden bridge across the Middle Saluda. Crossing a second, narrow log footbridge, you'll make a sharp right to start following a smaller tributary, Cox Camp Creek, on the trail up to Rainbow Falls.

About a mile from the trailhead, you'll cross a third wooden bridge, then pass through a slot between two giant boulders and turn right to start a steep ascent. At once you'll notice that the forest to your left is quite different from the one below and next to the river. For one thing, the canopy reveals that this is a drier slope, changing from an acidic cove forest to an oak forest. Canopy dominants here are northern red oak (*Quercus rubra*) and chestnut oak (*Q. montana*), as well as red maple (*Acer rubrum*). In addition, the understory is open, with fewer shrubs and saplings than along the river.

If you look closely at the bases of the trees, you'll notice that they are blackened, evidence of an illegal campfire that escaped and burned from campsite #1 in February 2009. The fire burned 350 acres; because it was late winter, the fire burned low, and it was not a hot burn. Its main effect was to kill the shrub and sapling layer of woody plants while leaving the canopy mostly intact. The result is a forest with a more open understory than you might expect.

Along this trail, the forest gets ever higher and drier. In one stretch after a major switchback to the left, large, gnarled chestnut oaks dominate the canopy, but they are joined by shortleaf pine (*Pinus echinata*), which makes up as much as 25% of the canopy cover in a few places. Here the understory (which is recovering from fire) favors the dry and acidic conditions, with mountain laurel (*Kalmia latifolia*), black gum (*Nyssa sylvatica*), and sourwood (*Oxydendrum arboreum*), plus several species of blueberry, all regrowing vigorously. One has to wonder if more frequent fires would favor pine renewal, while fire suppression would tip the balance toward oak dominance.

You will approach the base of a sheer wall of rock; this is Cleveland Cliff, right below YMCA Camp Greenville. The trail turns abruptly left and continues its climb. Notice the many stone steps along the trail, an impressive bit of engineering. The trail curves to the right one last time, and at this point you should be able to hear the roaring promise of Rainbow Falls. This protected ravine gets more and more species-rich as you approach the falls; look for abundant wildflowers during the growing season, and note that the canopy transitions back to dominance by cove forest species, including the first appearance of yellow buckeye (*Aesculus flava*), plus a cove regular you saw along the Middle Saluda, tulip-tree.

At 2.6 mi, you've reached your destination. Along the way you might have wondered, as you passed numerous tributaries, if each of them

When the Earth Moves

There is evidence throughout the Rainbow Falls hike of dynamic processes on the earth's surface, particularly waterfalls and rockslides. Waterfalls offer a a unique habitat (the spray cliff), home to plants that thrive on rock surfaces continually bathed in mist and spray from the cascading water nearby. On hot summer days, the microclimate is cool and moist in the vicinity of a waterfall, and during winter the constant humidity may ameliorate low temperature extremes. Winter brings icicles and ice masses that scour the rocks when they slip or fall, and flooding can scour the waterfall at any time of year, as can rocks that loosen and crash down. Many spray cliff specialists are poor competitors, however, and this constant disruption reduces competition from other species.

People love waterfalls, of course, and are attracted to many of the same conditions that favor the plants that grow there. They also come for aesthetic reasons, drawn to the ever-changing sights and sounds of cascading water, rainbows, and cool splash pools just begging for a dip. Waterfalls are especially abundant in the southeastern Blue Ridge Escarpment, owing to its high rainfall and rugged topography.

Rockslides, while lacking the aesthetic charms of waterfalls, are also critically important as habitat makers. A landslide can eradicate an old-growth forest, exposing rock and rubble to restart plant succession. Many fast-growing, short-lived species thrive in the sunny, resource-rich environments created by rockslides, and it is here that plants like blackberries (*Rubus* spp.), birches (*Betula* spp.), cherries (*Prunus* spp.), and others make their temporary homes.

We tend to take waterfalls and rockslides for granted as things that "happen" in the mountains, but this is far from true. Both phenomena indicate a landscape out of equilibrium. A waterfall occurs where there is a sudden drop in elevation. As waterfalls erode their way upstream, they leave below them steep-sided valleys prone to landslides. In this sense, waterfalls and rockslides share a common ancestry in the erosion caused by relentless water flow.

What causes sudden drops in topography and thus waterfalls? Continental drift can certainly create steep topography. However, the Appalachians haven't seen major tectonic activity for a couple of hundred million years, long enough for the erosive forces of water to smooth the landscape into one of rounded peaks and gentle valleys—and one lacking the excitement of waterfalls and rockslides.

Recent geological investigations have attempted to crack the mystery of why an ancient, eroded landscape like the southern Appalachians, which should be stable, has so many waterfalls and active rockslides. Geologists at North Carolina State University recently proposed that the southern Appalachians experienced "recent" uplift sometime in the Miocene—which ended about 5 million years ago—when a dense portion of the earth's crust broke off deep beneath the mountains. The loss of this dense material allowed the mountains to bob up (*very* slowly), like a swimmer letting go of something heavy. The newly uplifted surface features set the stage for rockslides and waterfalls. Waterfalls began migrating upstream in their newly renovated landscape. Meanwhile, downstream from these waterfalls, the resulting steep-sided valleys began again to experience the rumble of additional rockslides. These dynamic processes are still with us today, creating unique habitats and natural communities.

Next time you relax beside a waterfall or stumble through the rubble and dense vegetation of a rockslide, think about the renewing forces that keep the earth's surface young and dynamic, offering places for the diversity of living things to thrive. Waterfalls and rockslides are not just phenomena that "happen"; they are instead the surface expression of massive forces of renewal far beneath our feet. We truly inhabit a living earth, and when the earth moves, it sets in motion forces that provide us with some of the most amazing wonders we can experience on quiet mountain paths.

might be Rainbow Falls. But there is no mistaking Rainbow Falls—when you get there, you'll know it.

You will surely want to pause to enjoy the falls, maybe soaking your feet in one of the pools surrounding the cascade, before you make the much easier descent back to the trailhead. As you relax, examine nearby mossy rock crevices that are damp or dripping, looking carefully for salamanders. Children are especially good at spotting these fascinating creatures. A recent rainy visit yielded 5 species of the 16 that are found in the park. Salamanders breathe through their skin, so they thrive in places like these, which have ample damp niches for hiding. Some species are lungless, living their whole lives in water and

Rainbow Falls drops more than 100 feet, shattering into droplets as it breaks against the rocks below. On a sunny day, these tiny prisms of water make conditions ripe for a memorable rainbow.

breathing through external, feathery gills. Others are mostly terrestrial and live in the moist crevices between rocks. It's not surprising that this region—with its steep topography, mild winters, and high rainfall—has some of the highest salamander diversity in the world.

After your search for salamanders and a pause to enjoy the falls, follow the red blazes down the Rainbow Falls Trail until you cross the last wooden bridge and rejoin the blue-blazed Jones Gap Trail. As you saunter along the old toll road corridor on your way back to park headquarters, enjoy the forest of sweet birch, eastern hemlock (*Tsuga canadensis*), sycamore, beech (*Fagus grandifolia*), Fraser magnolia, red oak, red maple, and tulip-tree. This is an acidic cove forest, a fact made more evident by the heathy understory of rosebay rhododendron (*Rhododendron maximum*) and Carolina rhododendron (*R. carolinianum*), along with other shrubs tolerant of more acidic soils. As you hike through this seemingly stable forest, consider how catastrophic events like rockslides, fires, flooding, and clear-cutting alter forests suddenly, setting them on new trajectories that bear the imprint of the special challenges and opportunities presented to the flora.

11. SATULAH MOUNTAIN

Why Are We Here?

Satulah Mountain, owned by the Highlands-Cashiers Land Trust, is a great place to enjoy a short, easy hike that rewards you with a lunch stop on a rock outcrop with exceptional views of the Eastern Continental Divide, the southeastern Blue Ridge Escarpment, and (on a clear day) the Piedmont beyond. Here you'll encounter an excellent example of a montane oak forest and arrive at a granitic dome community at the trail's terminus. These are two important natural communities of the southern Appalachians. Along the way, the hike also offers an opportunity to see evidence of forest stratification and to meet a glacial relict.

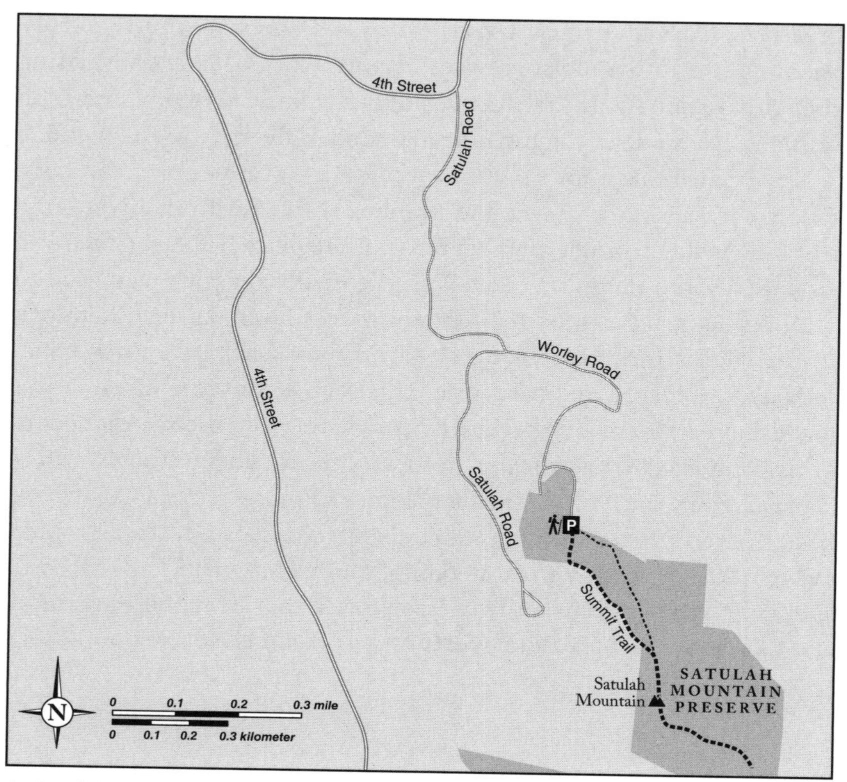

11. Satulah Mountain

The Hike

Our hike and our ecological journey both begin in the parking area. On the east side of the parking area is a rock face that was cut away when the parking area was established, creating a cross section of the community that occupies much of Satulah Mountain's ridge. This high-elevation red oak forest is one type of montane oak forest found at elevations above 4,000 ft throughout the southern Appalachians. Here we clearly see its profile and its prominent vertical strata, or layers. The canopy above, reaching about 30 ft, is dominated by northern red oak (*Quercus rubra*). Below the oaks are large mountain laurel (*Kalmia latifolia*) and occasional rosebay rhododendron (*Rhododendron maximum*), together forming a tall shrub layer to about 10 ft. Beneath these, galax (*Galax urceolata*) forms an herbaceous layer about a foot high.

11. SATULAH MOUNTAIN

Highlights: Gorgeous views of the Blue Ridge Escarpment into Georgia and South Carolina

Natural Communities: Montane oak forest, granitic dome

Elevation: 4,500 ft (trailhead); 4,543 ft (summit)

Distance: Less than 1 mi round-trip

Difficulty: Easy to moderate walking to the summit, with very little elevation change, but rocky in places. Use caution around the summit and rock outcrop areas, where there are steep drop-offs.

Directions: Satulah Mountain is just outside Highlands, NC, at the end of a continuously paved road. From the intersection of Main St. and US 64, head south on 4th St. (which is also NC 28). At 0.2 mi, negotiate the complicated intersection by jogging slightly left to continue uphill on Satulah Rd. (do not stay on NC 28, which turns right here). At 0.4 mi, bear left where Old Walhalla Rd. heads downslope to the right. At 0.7 mi, bear left on Worley Rd. At 1.0 mi, continue on the private road (do not take Vista Lane to the left). At 1.2 mi, pass a cell phone tower to your right and arrive shortly thereafter at the trailhead parking area (no facilities). The trailhead is at the south end of the parking area.

GPS: Lat. 35° 2′ 25.6″ N; Long. 83° 11′ 40.5″ W (trailhead)

Information: Highlands-Cashiers Land Trust, P.O. Box 1703, Highlands, NC 28741, (828) 526-9938

The structure of this community is easy to see, with three distinct layers. Ecologists are interested in stratification because of its importance in determining habitat suitability for animals, among other things. The diversity of various animal species increases with a corresponding increase in the vertical complexity of the vegetation. Why? The additional layers provide a greater variety of microhabitats that favor species with different needs.

As you begin your hike, you will notice several tree species in the canopy, in addition to red oak. These include other species of oak (white oak [*Quercus alba*] and scarlet oak [*Q. coccinea*]), along with serviceberry (*Amelanchier laevis*), sweet birch (*Betula lenta*), red maple (*Acer rubrum*), and on the drier parts of the ridge, pitch pine (*Pinus rigida*). Also in the canopy is a tree that you'll rarely see, the American chestnut (*Castanea dentata*), once an important component of this forest. Evidence of its former prominence can be found in the many American chestnut stump sprouts that you will see along the trail. Finally, look

Vertical stratification, or layering, is easy to see in this montane oak forest in the parking area at Satulah Mountain. The canopy has a mix of northern red and white oak, while mountain laurel and some rosebay rhododendron are found in the tall understory. The low ground cover shown here is galax.

for other interesting shrubs in the understory. In addition to mountain laurel and rosebay rhododendron, you may find the occasional mountain sweet pepperbush (*Clethra acuminata*), easily distinguished from the other shrubs by its peeling, cinnamon-colored bark.

As you hike along the ridge, pause to look for changes in canopy height and species composition as the trail takes you toward more exposed parts of the ridge. The canopy shrinks noticeably in exposed areas, likely reflecting the harsh winter conditions, when bitter winds drive ice and snow across the ridgetops. The ridgetops, with their shallow soils, dry out between rains faster than the adjacent slopes. These conditions favor pitch pine, which occurs in small clusters here and there. Occasional fires, like the one that swept up the ridge in 2009, help pitch pine to regenerate (see the sidebar "A Tale of Two Pines" in hike 22 [Table Rock]), and fire may be a key reason for its persistence in this community. Throughout most of the hike, though, you'll find yourself in a dense tunnel of evergreen shrubs, principally mountain laurel.

If you are fortunate enough to visit Satulah Mountain in late May through early June, you'll be rewarded by the stunning floral display of mountain laurel in full bloom. Take time to observe a mountain laurel flower closely; you'll see that each anther (the pollen-bearing part of the stamen) is tucked into a little pocket within the joined petals of the flower, ready to spring up to slap an unsuspecting pollinator with a load of pollen.

When you arrive at the summit, you'll see on your left a granite marker showing the elevation, 4,543 ft. As you emerge into the open, enjoy the panoramic view of the high country of South Carolina to your left (southeast) and upstate Georgia to your right (southwest). The most prominent peak you'll see from the summit of Satulah Mountain is Rabun Bald (in Georgia) to the south/southwest, in front of you and just to your right. You will also see several other dome-shaped rock outcrops along the Eastern Continental Divide in the immediate vicinity of Highlands. Beyond, the southeastern Blue Ridge Escarpment lies before you in a sweeping vista.

After enjoying the view from the summit, you'll want to continue walking along the ridge beyond. There is a network of braided trails that continues south toward a large, dry, south-facing rock outcrop, but stay to the main trail, which leads about a tenth of a mile to the rock outcrop. As you negotiate the sometimes winding trail to the outcrop, you may encounter a creeping shrub with needlelike leaves. This is ground juniper (*Juniperus communis* var. *depressa*), a glacial relict. We offer a brief discussion of glacial relicts, and some ideas about why they exist, in the sidebar. The trail then takes you to a spur that continues to the rock outcrop. Here again you'll find spectacular views—bring your lunch to enjoy along with the views, but take care with your footing!

While you're having lunch on the rock outcrop, you may wonder why such large, open outcrops exist in the first place. In the case of Satulah Mountain and several other similar dome-shaped outcrops in the southern Blue Ridge, the answer lies buried (literally and figuratively) deep in geologic history. Most of the peaks of the southern Appalachians are highly eroded remnants of a mighty mountain range

The open summit of Satulah Mountain provides niches for glacial relicts and offers sweeping views of Rabun Bald and the Blue Ridge Escarpment.

that rivaled the present-day Himalayas. Between 450 and 250 million years ago, the continental plates of ancestral Africa (Gondwana) and North America (Laurentia) came together in a series of collisions that formed the southern Appalachians as part of a single supercontinent, Pangea. In some of these earliest mountain-building periods, molten rock, or magma, welled up beneath weaker areas in the earth's surface, forming large "domes" beneath the existing rocks. This magma never broke through the surface rock to form volcanoes but simply cooled in place, forming granite when it hardened.

Over time, the softer surface-rock layers weathered away, revealing the harder and more erosion-resistant granite. Eventually, the domes emerged as prominent peaks towering above the surrounding landscape. Freeze/thaw cycles result in fracturing and sloughing of layers of rock parallel to the surfaces of the domes, revealing more hard rock beneath. The smooth surfaces of granitic domes lack fissures or crevices where soil can accumulate, so the steeper parts of the domes remain sparsely vegetated.

Plants found on these granitic domes must be capable of growing

Ice Age Relicts

Ground juniper (*Juniperus communis*) is a rare glacial relict found around high-elevation, dry rock outcrops in the southern Appalachians, but it is far more common to the north. The term "glacial relict" means that this species is thought to have been more widespread in the South during times of cooler climate, but it is now found only rarely in higher elevations. As the climate warmed at the end of the most recent Ice Age, many cold-adapted plants and animals retreated northward in the wake of retreating glaciers, and many species went locally extinct in the South. Other species succumbed to the changing climate altogether. The few cold-adapted species that persist today, the glacial relicts, do so in places that afford cooler microclimates and open habitats free from competition from overtopping trees, such as the granitic dome of Satulah Mountain.

To place the significance of these rare rock outcrop plants in context, imagine emerging from the rhodendendron tunnel of the Satulah Mountain Trail and encountering a woolly mammoth (*Mammuthus primigenius*) in the summit mists. The story (and shaky cell-phone photo) of your encounter with this prodigious pachyderm would no doubt make headlines across the globe. In a sense, this meeting would be no less remarkable than finding the suite of glacial-relict rare plants that occur on Satulah Mountain, Craggy Pinnacle, and other rock outcrops today. Although the mammoth apparently succumbed to a lethal postglacial mix of climate change and Neolithic hunters, the many rare plants that persist here on Satulah and elsewhere still maintain their tenuous grip on the rocks they call home.

on bare rock (such as some lichens and mosses) or be capable of eking out an existence from the minimal resources (water and nutrients) available in thin patches of soil that accumulate on small ledges or in surface depressions in the rock. Only sunlight is available in abundant supply, favoring plants requiring high light levels but also contributing to the dry nature of this rigorous habitat. The "Extreme Botany" sidebar in hike 25 (Grandfather Mountain) offers additional thoughts about adaptations of species that survive in extreme environments.

Look around and appreciate how drastically foot traffic can affect these tenacious, yet fragile, communities. To get a good sense of the granitic dome community, walk over to a less-trafficked area where the rock is open yet vegetated with herbaceous plants and low-growing shrubs around the edges. You will notice immediately that microtopography plays a key role in the granitic dome natural community, since even the smallest depressions accumulate soil and funnel rainwater. There is, in fact, a correlation in these communities between soil depth and the type of plant cover. The smallest plants, such as lichens, cling to the bare rock, and increasingly larger plants grow in the deeper pockets of soil.

Aside from lichens and mosses, you will observe several herbaceous plants in the granitic dome natural community. Two of them are endemics, found only in the southern Appalachians: mountain dwarf-dandelion (*Krigia montana*), with its nodding yellow dandelionlike flowers scattered here and there throughout the summer, and a primitive plant, twisted-hair spikemoss (*Selaginella tortipila*). Several other herbaceous species, including various grasses (such as mountain oat grass [*Danthonia compressa*]), can also be found here. Fringing the open rocks where soils are a bit deeper are various shrubs, including some you observed on the hike to Satulah's summit, such as mountain laurel. Among the shrubs, a rock outcrop specialist is sand myrtle (*Kalmia buxifolia*), which forms a low fringe around the more open rock faces. Its scientific name implies that the leaves are reminiscent of those of the ornamental boxwood, although much smaller.

What keeps these granitic domes open and treeless? You might expect that, over time, the rock would continue to weather and soil would continue to accumulate in pockets. If the soil depth continued to increase, larger shrubs and even small trees could take root, eventually blanketing the summit with a montane oak forest similar to the one you observed on your hike up. However, these summits typically remain open, because of the combined forces of continued sloughing of the granitic dome as well as harsh weather conditions, including wind, snow, ice, and frequent summer thunderstorms with accompanying deluges of rain. These disturbances tear the taller, exposed vegetation off the mountaintop, so the smaller plants (and panoramic views) remain intact on this summit.

When you finish exploring the summit and adjacent rock outcrop, return to the parking area via the same trail (another option is to take the newly opened trail to your right just below the summit, which takes you out to additional views). On your descent, pay close attention to the structure of the forest as you move from the extreme, windswept, and sparsely vegetated summit to the short-statured forest and eventually into the additional layers and complexity of the taller forest near the parking area.

---

12. WHITESIDE MOUNTAIN

Why Are We Here?

Whiteside Mountain is home to several different kinds of plant communities, primarily because environmental factors change as the trail winds its way around the summit. We will study two important second-growth forests, which developed after moderate disturbance, as well as two plant communities that occupy rock outcrops.

Most visitors to the southern Appalachians probably think that the lush and extensive forests they see have been there forever. With few exceptions, however, nothing could be further from the truth. You will see and interpret the effects of past logging and examine the environmental factors that result in different plant communities.

The Hike

From the US Forest Service kiosk clearly visible from the parking area, ascend a short stairway and turn left onto the main trail, which follows an old logging road. Hike about 0.1 mi along this road until you reach the intersection with the Whiteside Mountain loop trail. To your right, the loop trail ascends steeply. Straight ahead, the loop trail continues to follow the old road, and we recommend continuing your hike along this road, beginning a clockwise hike around the loop. This strategy offers a gentler hike to the summit, with a return to this intersection on the steeper trail to your right.

As you resume your hike, you'll be walking through a lush forest that covers the generally northwest-facing slope of Whiteside Mountain. A few hundred feet after passing the loop trail intersection, care-

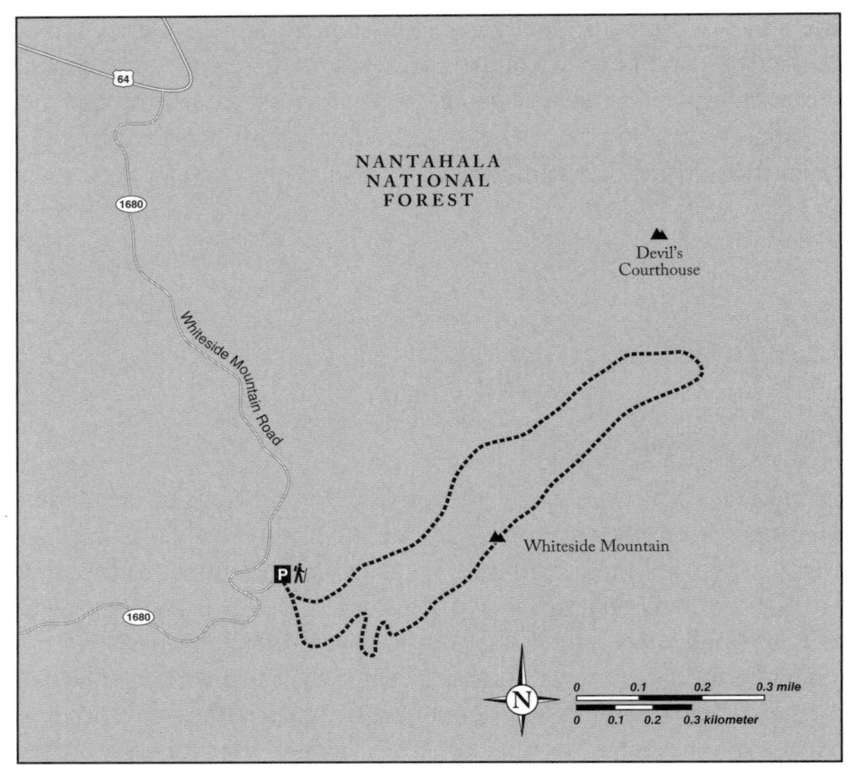

12. Whiteside Mountain

fully examine the 10-ft-high vertical rock face to your right. If you look closely, you can see in cross section the many holes that were drilled into the rock during road construction to facilitate blasting. Clearly, this road to the summit was built at great effort and expense. Why? According to the US Forest Service, Whiteside was purchased by a corporation in the early 1900s as a tourist attraction, and shuttle buses once carried visitors to the summit.

After a decline in tourist traffic during the Great Depression, the mountain's timber was sold and logged in the 1940s, before the mountain was acquired by the US Forest Service. As you walk this quiet pathway today, imagine the heavy logging trucks hauling trees cut from this forest to the nearest mill. You'll then realize that the forest you see today is what ecologists call a secondary or second-growth forest, one that regrew naturally after timber harvest, described in more detail in the sidebar.

12. WHITESIDE MOUNTAIN

Highlights: Sweeping Blue Ridge views, colorful summer wildflowers, and a hanging garden in a seepage wetland

Natural Communities: Northern hardwood forest, seepage wetland, montane oak forest, granitic dome

Elevation: 4,440 ft (trailhead); 4,930 ft (summit)

Distance: 2-mi loop trail, round-trip

Difficulty: Moderate climb to summit, easy walking on fire roads, some steep descents

Directions: Whiteside Mountain is located off US 64 between Highlands and Cashiers. From Highlands, take US 64E about 5.5 mi until you see a sign directing you to Whiteside Mountain. Turn right on SR 1680, then drive about 1 mi to the US Forest Service parking lot on the left (trailhead, vault toilets, parking fee, picnic tables).

GPS: Lat. 35° 4′ 49.4″ N; Long. 83° 8′ 39.2″ W (trailhead)

Information: Nantahala National Forest, Nantahala Ranger District, 90 Sloan Rd., Franklin, NC 28734, (828) 524-6441

The forest community along this section of trail is an example of a northern hardwood forest. Such forests contain hardwood species usually associated with more northern climates, and, indeed, species of yellow and sweet birch (*Betula alleghaniensis* and *B. lenta*) are common in the canopy, interspersed with cucumber magnolia (*Magnolia acuminata*), Fraser magnolia (*M. fraseri*), black cherry (*Prunus serotina*), and red maple (*Acer rubrum*). Notable species in the subcanopy include a smaller tree, striped maple (*A. pensylvanicum*), and an abundant tall shrub, witch hazel (*Hamamelis virginiana*).

Pause and examine the 10 largest trees within your range of sight. How many of them are northern red oaks (*Quercus rubra*)? What smaller plants are growing near the ground in the understory? Please keep your answers in mind as you continue to ascend—you'll revisit these questions near the summit.

The northern hardwood forest community at Whiteside Mountain grows over soils formed by the weathering of the ancient granite that forms the dome. Because of the steep slopes, the soils here are thin and acidic. The understory plant community reveals this fact without soil testing. You might see a few pockets of ferns, sedges, and wildflowers, but most of the forest beside the trail grows thick with members of the acid-loving heath or blueberry family, including three species of

The old road and even-aged stand of northern hardwood trees are clues to a past history of logging and a second-growth forest stand at Whiteside Mountain.

evergreen rhododendrons (the common, white-flowered rosebay [*Rhododendron maximum*], the similar-looking but pink-flowered Catawba [*R. catawbiense*], and the smaller-leaved and purple-flowered gorge [*R. minus*]), along with mountain laurel (*Kalmia latifolia*) and several other species from the family.

As you continue toward the summit, on your right you will pass several rock faces of varying widths (30 to 100 ft); these open faces are usually damp if not actually dripping. Here you'll discover the seepage wetland community. What sorts of plants do you see growing on these wet rock faces? Does the vegetation change as the steepness of the rock changes? On the steepest faces, only a carpet of moss and a few lichens cling to the slippery surface. In small pockets and crevices that have accumulated soil, you might see a few larger plants. And you'll notice that few, if any, trees can find enough soil to anchor their roots.

Plants growing on these northwest-facing seeps must be adapted to continual saturation and very thin soils. Although few species can live here, one that you'll likely see is deerhair bulrush (*Trichophorum*

Second-Growth Forests

Marks where the rocks were drilled to insert dynamite are sure signs of heavy-duty human activity on the flanks of Whiteside Mountain. From these marks, the presence of a substantial logging road, and US Forest Service records, we can learn that this area was logged in the mid-20th century. If you had only the clues provided by the forest itself, would you know that you are walking through a secondary forest? If you study the forest that slopes to the left of the trail, below the old logging road, you'll notice that there are few of the large, old trees that characterize old-growth forests. The occasional very large tree is likely to be crooked or otherwise unsuitable for milling, left behind as loggers extracted the high-quality timber from this site. Most of the trees filling the canopy today grew here after the site was logged and are therefore of a similar size (about 1 ft in diameter) and age.

Ecologists and foresters would refer to this stand as "even-aged," in contrast to the all-aged forest that one encounters in old-growth. Look for other evidence of past logging, such as old cut stumps and multistemmed trees that originated as sprouts from some of the cut stumps. These additional clues should reinforce your hypothesis that this is a second-growth forest.

You may also notice many standing but recently dead trees. These are fire cherries (*Prunus pensylvanica*). These dead fire cherries are another clue to the past logging history; this species rapidly colonizes higher elevations in the southern Appalachians following logging, but it is relatively short-lived. Another early colonist of cut-over forests is black locust (*Robinia pseudoacacia*), notable for its compound leaves and rough, furrowed bark. While fire cherry captures nitrogen released into the soil after harvest, black locust fixes atmospheric nitrogen into a form that can be taken up by plant roots. Both contribute to the soil's fertility after severe disturbance, paving the way for a healthy, regenerating northern hardwood forest.

We have highlighted these interpretive clues to past forest history in the context of Whiteside Mountain, but we hope you will be able to apply these to other settings in the southern Appalachians. While the specific players may vary from place to place—you won't find fire cherry at lower elevations, for example—the general principles of seeing the forest with the trees remain the same.

cespitosum). Scientists call this a glacial relict species, as it is far more common to the north. It is believed to have had a wider distribution in this area during periods of cooler climate, particularly during the last glacial period, when alpine tundra occupied the higher elevations in the southern Appalachians.

Today, forests cover the southern Appalachians to the highest summits, so species needing open habitats are limited to small areas, such as rock outcrops, where trees can't get a foothold. Look on the wet rock faces for the tiny, white, starry flowers and coarsely toothed rosettes of basal leaves of Michaux's saxifrage (*Micranthes petiolaris*) in early summer. Granite dome St. John's-wort (*Hypericum buckleyi*) is also present in rounded clumps of bright green foliage topped with brilliant yellow flowers with showy stamens.

Above the seepy rock faces, enough soil has accumulated in places to support other, larger plants, while shrubs and trees hug the edges of the openings. In July, you'll see the nodding orange heads of the majestic Turk's cap lily (aptly named *Lilium superbum*) perched on the occasional small ledges that occur on the seepy rock face. On the larger seepy rock faces, you will have a clear view up the slope toward the summit. What keeps these places open, when dense patches of rhododendron grow just 10 ft away on each side? The answer may be found in the winter months. If you are visiting then, you will discover these rock faces covered with thick sheets of ice. Thawing in early spring sends large chunks of ice tumbling downslope, removing mats of vegetation and soil in their path. Thus these seepage areas are dynamic habitats, subject to disturbance, resulting in a habitat nearly devoid of trees or shrubs.

As you continue climbing, especially above 4,800 ft, you'll notice a gradual change in the forest's composition, which becomes increasingly dominated by northern red oak. Eventually, northern red oak becomes sufficiently dominant that we can no longer call this community a northern hardwood forest but, rather, a montane oak forest, one of several kinds of oak forest found in the southern Appalachians.

About 100 yards before you reach the first open vista where the trail turns to the right to follow the summit ridge, stop again and observe the 10 largest trees within your sight, noting how many of these are northern red oak. What do you discover when you compare this number to your earlier count? What does the understory look like here?

Can you see how the forest canopy has changed from mixed northern hardwoods to dominance by northern red oak?

To be sure, northern red oak is an important component of the northern hardwood forest, but here other northern hardwood species, such as yellow birch, are reduced in abundance. The understory, too, changes its character, with evergreen rhododendrons becoming less common, replaced by a more open understory dominated by ferns, sedges, and many species of wildflower, particularly members of the aster family. Why this gradual shift in forest composition? The increasingly cooler climate associated with higher elevation may play a role here, but the slopes along this part of the trail are also gentler. Greater depth of soil accumulation on these gentler slopes may provide some advantage to herbaceous understory species, which replace the evergreen rhododendrons that typically dominate on steeper slopes with shallower soils.

The gradual transition from one forest community to another (in this case, northern hardwood forest to montane oak forest) happens whenever environmental conditions also undergo gradual change. As you sharpen your ability to read forest communities, watch for these subtle transitions, which can make it difficult to place a forest neatly into a particular community type. If you enjoy thinking about this subject, you will find the sidebar "Community Concepts in Conflict" in hike 15 (Gregory Bald) of interest.

When you finally reach the first open, shrubby area near the summit, pause to enjoy spectacular views of the Eastern Continental Divide in front of you and your first sweeping view of the headwaters of the Chattooga River to the right in the valley far below. The trail then turns sharply right and continues along the summit ridge. As you turn to start this part of the trail, you can descend a few steps and take in the panorama from an overlook platform.

Along this stretch of trail, you will find many openings that afford spectacular views of the Blue Ridge Escarpment. These open areas contain granitic dome communities, comprised of treeless expanses of bare rock with few plant species. While hiking the summit ridge, take care to stay on the marked trails. This is for your safety as well as that of some of the fragile plants that occupy the rock outcrops beyond the barriers. Compare the granitic dome communities on the trail side of the barrier with those in the section protected from hikers by a fence, and realize just how easily they can be obliterated by foot traffic. As the sidebar "Loving a Place to Death" in hike 19 (Craggy Pinnacle) discusses, human trampling of rare plants is a serious problem in the region.

Any of the dry, rocky outcrops to your left along the summit trail are good places to observe the granitic dome community and compare it with the seepage wetland you studied on your hike up the old logging road. The environmental factors that dry and wet rock outcrops have in common are steep slopes and thin to nonexistent soils. In fact, you will see a few of the same species that you observed on the seepage area—namely, Michaux's saxifrage and granite dome St. John's-wort.

However, granitic dome and seepage wetland communities differ greatly with respect to moisture. The south-facing slopes at Whiteside are exposed to intense sunlight all summer, and this factor, combined with desiccating winds and thin soils, makes living here difficult for most plants. One of the most abundant plants is the mosslike twisted-hair spikemoss (*Selaginella tortipila*), which forms dense mats along rock crevices and in pockets of thin, rocky soil. Visitors in early June will be treated to a flowering display of rock harlequin (*Capnoides sempervirens*), with its yellow and coral two-toned flowers.

Keep walking and you'll reach the actual summit just beyond the sign that tells a story about a heroic rescue at Fool's Rock. While walking along the summit ridge, pause to reexamine the high-elevation red oak forest to your right. Although the canopy oaks are stunted and gnarly because of the harsh winter conditions here, the natural community will be quite familiar to you, because the northern-red-oak-dominated forest you observed on the hike up extends to the crest of the ridge. Only when soils become too shallow to support trees do the oaks yield to the predominantly herbaceous and nonvascular plants that comprise the granitic dome community.

Past the summit, you will begin to descend, and you will soon reach a stairway, where you'll pass another seepage wetland to your right. See how many of the now-familiar plants you can identify! Turk's cap lily, deerhair bulrush, Michaux's saxifrage, and granite dome St. John's-wort are all present here during the growing season. Watch your footing on the steep rocky areas after you leave the steps; these are often wet and slippery. You will soon reach the intersection of the loop trail and the old logging road where you began this loop hike, and it is only a short walk back to the parking area to complete your hike.

13. HOOPER BALD

Why Are We Here?

The Hooper Bald Trail provides easy access to a grass bald, an uncommon southern Appalachian community type. Grass balds are treeless (or nearly treeless) areas typically found between 3,500 and 5,500 ft in elevation and dominated by grasses, herbaceous plants, and scattered shrubs. They are popular areas for hikers and picnickers because they offer great views on clear days, berry-picking in season, and spectacular displays of flame azaleas (*Rhododendron calendulaceum*) in June.

The hike to Hooper Bald will take you through a short-statured northern hardwood forest that surrounds the bald. Such a forest is common at this elevation, leading us to wonder why the bald exists at all. Your study of the bald and surrounding forest will reveal some interesting stories about effects of human influence on the landscape. This trail abounds with evidence of human activity and other contributing factors, which suggest the origins and continuance of these fascinating and beautiful open areas in the mountains.

An important historical note is that Hooper Bald is the introduction point for the exotic European wild boar (*Sus scrofa*) in the southern Appalachians. This animal has since established itself in the region and caused considerable damage to plant communities of high elevations, most notably by uprooting and decimating spring wildflowers. George Moore imported the boars into a large, fenced game reserve established at Hooper Bald. Over time, some animals escaped from the enclosure, and the reserve itself was eventually shut down. Of the big game animals housed in Moore's reserve, only the boars thrived outside captivity and spread throughout the southern part of the southern Appalachians.

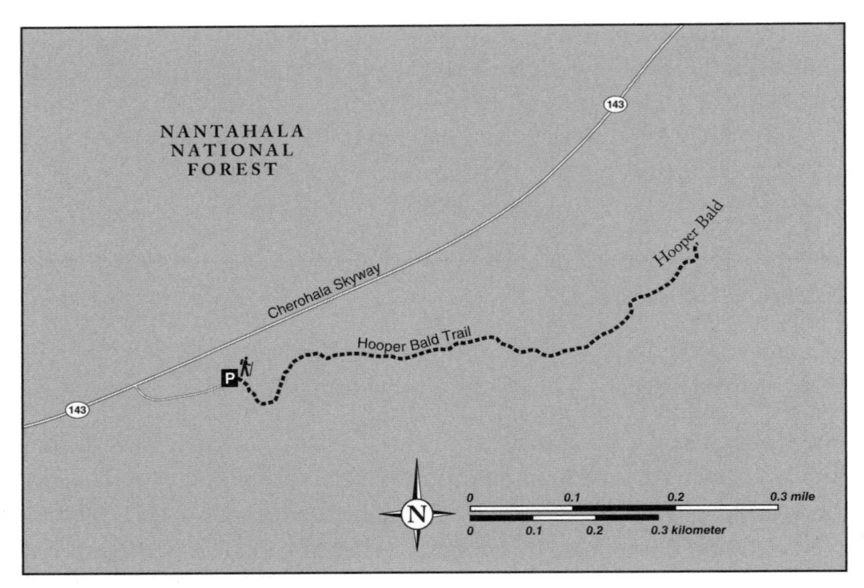

13. Hooper Bald

The Hike

The trail, a relatively level gravel path, begins at the parking lot and picnic area and traverses the northern hardwood forest, reaching Hooper Bald in just under half a mile. As you hike through this attractive forest, note the key species in the canopy.

Yellow birch (*Betula alleghaniensis*), American beech (*Fagus grandifolia*), and northern red oak (*Quercus rubra*) are the canopy species here, with buckeye (*Aesculus flava*) and sugar maple (*Acer saccharum*) represented to a lesser degree. Take note that these species differ significantly in their size and representation in the forest—this disparity will help you construct a story about the forest's origins.

Immediately after crossing a small dirt access road with a gate, you'll see some very large trees (24+ inches in diameter) surrounded by a dense forest of much younger trees. Notice that the large trees are almost exclusively northern red oaks. Choose one of these to study and note that, in addition to its large diameter, the tree has a very large, spreading crown. Typically, forest trees have narrow crowns that are crowded by the crowns of other trees competing to capture the available sunlight in the closed forest. The spreading crowns of

13. HOOPER BALD

Highlights: Mysteries of grass bald origins; wild strawberries and blueberries in season; beautiful vistas; great picnic spot

Natural Communities: Northern hardwood forest, grass bald

Elevation: 5,290 ft

Distance: 1 mi round-trip

Difficulty: Easy walking on a mostly gravel pathway

Directions: From Tellico Plains, TN, follow signs to head east on TN 165 (the Cherohala Skyway). The Hooper Bald parking area is on your right, 7.6 mi past the NC/TN state line (picnic tables, vault toilets).

From Robbinsville, NC, follow Atoah Rd. west, then turn left to continue on Atoah Rd., which merges with NC 1127, Snowbird Rd. Follow Snowbird Rd. for about 7.7 mi. The road turns slightly left and becomes the Cherohala Skyway (there are signs). Continue 9.7 mi; Hooper Bald will be on your left.

GPS: Lat. 35° 18′ 15.1″ N; Long. 84° 0′ 3.9″ W (parking area)

Information: Cherohala Skyway Visitor Center (TN), 225 Cherohala Skyway, Tellico Plains, TN 37385, (423) 253-8010

Cherokee National Forest (TN), Tellico Ranger Station, (423) 253-2520

Nantahala National Forest (NC), Cheoah Ranger Station, (828) 479-6431

these oaks tell you that these trees grew up in an open area, likely an area that looked like the adjacent grass bald.

Ecologists sometimes refer to stands of high-elevation red oak like this one as "oak orchards" because of the resemblance of the spreading red oaks to ancient apple trees. These large red oaks are surrounded by much smaller beech and yellow birch that are all about the same size, about 6 to 8 inches in diameter. Many of these smaller (and younger) trees have multiple stems; some come from the root stock, and others branch about 1 ft above the ground.

Clearly, there was a substantial time gap between the establishment of the large northern red oaks and the smaller yellow birch and American beech. We hypothesize that this age disparity reveals that the grass bald once covered a greater area of this mountaintop and that it was likely maintained by grazing of domestic livestock. Red oaks were kept as shade trees in this pastoral setting, and these trees developed the large, spreading crowns typical of trees grown in

The large, spreading canopy of a northern red oak in the background suggests that the tree spent its formative youth in the grazed, open grass bald, while the younger, straight-trunked yellow birches in the foreground clearly grew up inside a closed-canopy forest more recently.

the open. Over time, the pasture was abandoned, and yellow birch and beech established from both seed and sprouts on the site. Soon after the abandonment, it seems that some effort was made to cut these stems and return the developing forest back to an open field. This explanation would account for the large number of yellow birch and beech that have trunks that are split about a foot above the root crown into multistemmed trees.

Interestingly, none of the young saplings or tree seedlings here are northern red oak. We suspect that red oak could not establish well in the fast-growing, competitive environment that existed after grazing stopped. Conditions now are apparently challenging for red oak regeneration; the red oaks probably can't compete for root space and sunlight in the dense thickets of yellow birch and American beech. Thus, despite an ample acorn crop produced by the large, mature red oaks, we don't see oak regeneration. Instead, the understory contains sugar maple, flame azalea, highbush blueberry (*Vaccinium corymbosum*), hawthorn (*Crataegus* sp.), withe-rod (*Viburnum cassinoides*), and some higher-elevation forest herbs.

In some places, the lush herb layer is diverse, with many showy species, such as the beautiful tassel-rue (*Trautvetteria caroliniensis*). In other areas, a single species of sedge (*Carex pensylvanica*) dominates the herbaceous layer; its dense sward of linear leaves appears as if it had been combed by a dedicated but unseen forest groomer!

Just past a half mile from the trailhead, you'll reach the opening onto the grass bald. You can hike the trail to the right less than a hundred yards to discover a nice picnic spot on a rock outcrop to your right, or you can continue past the outcrop through the center of Hooper Bald for great views of the surrounding mountains. A visit in mid-June will yield several treats, among them a view of flame azaleas along the edge of the bald. Also, look down at your feet for the toothy, trifoliate leaves of wild strawberry (*Fragaria virginiana*). The fruits are small but incredibly sweet. Be sure to allow time for berry-picking in season.

You'll also notice that a large area of the bald is covered with invading shrubs. A mid-July visitor to Hooper Bald might discover the tasty highbush blueberry in fruit, particularly around the edges of the bald. Nearly all of the grasses you see in the bald are of European origin, with the exception of sparse populations of mountain oat grass (*Danthonia compressa*). Why might this be the case? What can these clues tell us about grass bald origins?

Grass balds *are* a bit of a mystery; scientists are still uncertain of their origins and what forces have kept these places open and treeless. We've already suggested that Hooper Bald was formerly kept open by grazing of domestic livestock, but why was it open to begin with?

One hypothesis that has since been dismissed is that Native Americans cleared and burned these places to use as lookouts and hunting grounds. Other stories are more imaginative. At the trailhead, an interpretive sign shares a Cherokee legend describing a giant, green-winged hornet called Ulagu that carried off young children, and how the Great Spirit had split open the mountain so the Cherokee warriors could slay the beast. Subsequently, the Great Spirit rewarded the Cherokee by keeping the balds open for hunting and lookouts. However, some Native American legends such as this one have recent origins and were concocted to fool European settlers. Furthermore, few

Grass Bald Origins and Future

Alan Mark of Duke University first hypothesized that grassy balds originated when climatic shifts pushed trees off the highest mountain summits (most recently during the recent Ice Age, at its maximum 18,000 years ago), leaving them treeless. It was further developed by Peter Weigl and Travis Knowles, of Wake Forest and Francis Marion Universities, respectively, who suggested that these alpine meadows were then kept open first by large native grazers, collectively called megafauna (mammoth, mastodon, bison, horse, tapir, musk ox, ground sloth, elk, and deer), then more recently by domestic grazing animals brought by European settlers.

We know from historical documents and journals that when European settlers arrived, the grass balds already existed. With domestic animals, European settlers also introduced weedy European grasses. Grass balds today are typically dominated by European weeds and have few native plants. The vast majority of native plants found on balds are common elsewhere, suggesting that the balds originated recently enough to lack a distinctive flora.

The Weigl-Knowles hypothesis, sometimes called the Pleistocene megafauna hypothesis, has some romantic appeal: standing in the center of Hooper Bald, with its 360° views of the mountains, can't you just imagine woolly mammoths (*Mammuthus primigenius*) lumbering over the grassy slopes?!

Regardless of the origins of grass balds, one thing is certain: whatever forces created the balds and kept them open in the past are no longer operating. These treeless expanses are gradually closing in from the edges with shrubs, such as highbush blueberry (*Vaccinium corymbosum*) and the gorgeous flame azalea (*Rhododendron calendulaceum*), followed by high-elevation species from the adjoining forests. Various management agencies have debated the merits of keeping these balds open, knowing that they may in fact be human artifacts and not natural communities. However, agencies such as the National Park Service and US Forest Service recognize the historic and scenic value of these unique places and compromise by keeping a select few of them open. In Great Smoky Mountains National Park, Gregory Bald and Andrews Bald are maintained to remain open; the US Forest Service likewise maintains Hooper Bald through mowing and hand-cutting. The rate of change from grass to shrub cover is relatively slow, since few birds visit the grassy area to disperse seeds. Thus it is possible to keep the balds open with a modest investment in annual maintenance.

Regardless of your views on management of communities that may not be entirely natural (existing independent of human activities), you'll have to agree that grass balds like Hooper Bald are beautiful places, destinations for humans and wildlife alike.

Native American artifacts have been found by archaeologists on balds such as Hooper Bald. Without stronger evidence for Native American establishment, we must abandon this hypothesis and look for other explanations, as discussed in the sidebar.

Allow yourself some time to enjoy the views before turning around and returning to the parking area via the same trail; on your return trip, enjoy the vistas and pick some berries while pondering the origins of these high and beautiful places.

14. JOYCE KILMER MEMORIAL FOREST

Why Are We Here?

Joyce Kilmer Memorial Forest is a "must-do" hike for anyone interested in experiencing old-growth forests in eastern North America. This is one of the best-known and most beloved of all the remaining old-growth forests in this part of the world. Here you will encounter outstanding examples of both acid and rich subtypes of the cove forest natural community, both containing large, old trees of enormous proportions. These forest giants inspire awe, but they are just one component of the old-growth forest. During your hike, you will learn that an old-growth cove forest is much more than a stand of very big trees.

As you hike through this beautiful stand, you'll no doubt wonder

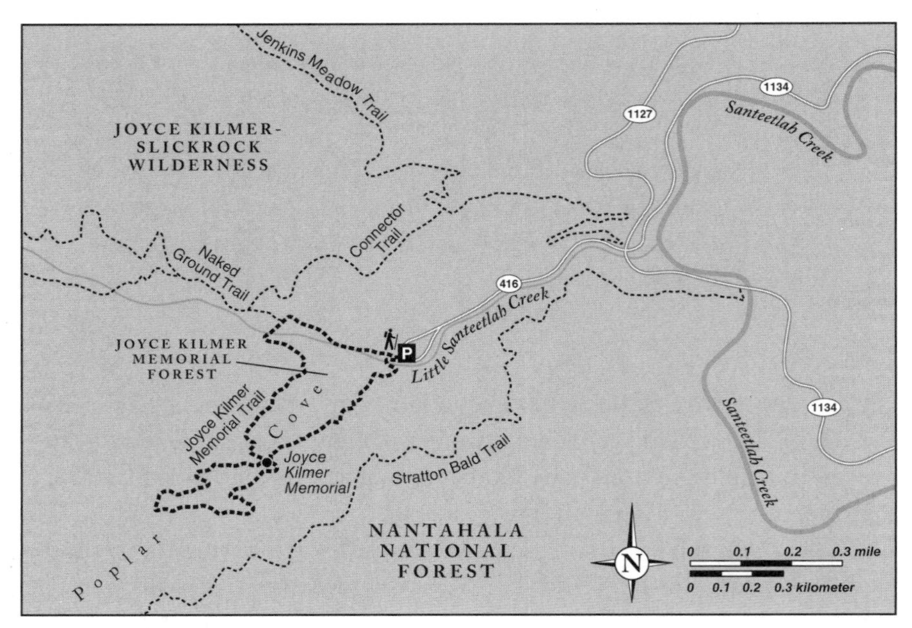

14. *Joyce Kilmer Memorial Forest*

how it came to be and how it is changing. We can speculate about the forest's origin, and ecologists understand much about its current dynamics, thanks to several studies of forests like this one. Because the forest contains a substantial component of eastern hemlock (*Tsuga canadensis*), it, like many other forests in the southern Appalachians, is undergoing dramatic change resulting from mortality caused by the exotic insect hemlock woolly adelgid (*Adelges tsugae*).

During your hike, you will also encounter this forest's diverse flora, in terms of both canopy trees and, in the rich cove forest subtype, the spectacularly diverse display of wildflowers. Understanding the causes of this diversity will lead us to examine the very early origins of forests like this one, which began during the Age of Reptiles. There we'll find important clues as to why these forests look like (and share many plant genera with) forests of eastern Asia.

The Hike

What is now Joyce Kilmer Memorial Forest was part of a privately owned tract scheduled to be logged in the 1930s. Fortuitously, increas-

14. JOYCE KILMER MEMORIAL FOREST

Highlights: Cathedral-like old-growth forest, with giant tulip-trees and gorgeous spring wildflowers

Natural Communities: Cove forest (both acidic and rich subtypes)

Elevation: 2,240 ft

Distance: 1.25-mi lower loop trail connects to 0.75-mi upper loop for a combined distance of 2 mi round-trip

Difficulty: A moderate climb at the trailhead, then easy walking on level ground around the upper and lower loops

Directions: From Robbinsville, take Hwy 129N for 1.5 mi to the junction with Hwy 143W (Massey Branch Rd.). Turn left and proceed west for approximately 5 mi to a stop sign. Turn right onto Kilmer Rd. Drive 7.3 mi and arrive at the top of Santeetlah Gap and the junction with the Cherohala Skyway. Bear right and continue for another 2.5 mi to the entrance of Joyce Kilmer Memorial Forest. Turn left into the entrance; it's about a half mile to the large parking area (sheltered information kiosk, picnic tables, restrooms). The trailhead is at the far end of the parking area near the information kiosk.

GPS: Lat. 35° 21′ 32.2″ N; Long. 83° 55′ 46.2″ W (parking area)

Information: Cheoah Ranger District, 1070 Massey Branch Rd., Robbinsville, NC 28771, (828) 479-6431

ing industrial demand for aluminum prompted expanded processing of the abundant bauxite ore in the region, a technology that required enormous amounts of electric power. Numerous rivers were dammed for power generation, and one of the resulting lakes (Lake Santeetlah) blocked rail access to this tract just prior to its proposed logging. After the 1929 stock market crash, timber prices plummeted and the US Forest Service acquired the land, including 3,800 acres of virgin (never-logged) forest. A portion of this magnificent forest was set aside as a living memorial to Joyce Kilmer, an American soldier from New Jersey who was killed while fighting in France during World War I. Kilmer is famous for his poem "Trees," so it is fitting that this memorial contains some of the largest remaining trees in the southern Appalachians. The memorial was dedicated on July 30, 1936.

The hike consists of two intersecting loop trails forming a figure 8. The lower of these is the 1.25-mi Memorial Loop, accessed by a short trail that takes the hiker from the visitor kiosk across Little Santeetlah Creek and to the base of the Memorial Loop, orienting the hiker in a counterclockwise direction. The relatively steep ascent from the creek traverses a good example of acidic cove forest. The canopy here is dominated by eastern hemlock, and the understory is densely shrubby, with rosebay rhododendron (*Rhododendron maximum*) dominant.

In recent years, the old-growth eastern hemlock trees have died as a result of the hemlock woolly adelgid and have begun to fall, leaving what was once a deep forest in stark sunshine, with dense shrubs competing for the light now available. To protect visitors from the hazard of falling trees, the US Forest Service took down some of the skeletal giants using dynamite in 2010. The reason is twofold: first, because this is part of a federally designated wilderness, no mechanized equipment is allowed, and second, because dynamite gives the snags left behind a more natural appearance than sawn stumps. The story of how the mighty eastern hemlock is succumbing to attack by a tiny exotic invasive insect is told in the sidebar "Hemlock Woolly Adelgid" in hike 18 (Ramsey Cascades).

As the trail levels out, the natural community shifts to a rich cove forest, with an open understory that provides a better view of the forest. The terrain is more gently sloping here, so this is a good place to start a list of the trees you can identify, dividing it into those you see in the canopy and those that are growing beneath the roof of the forest. As you continue upward toward the memorial plaque (affixed to a large boulder) that marks the intersection of the two loop trails, you'll be surprised at how many different kinds of trees you have on your list.

The area surrounding the memorial plaque is an excellent rest stop and a good place to reflect on what you have just seen. You have no doubt noticed that there are trees of all sizes in this forest, from small seedlings and saplings through mid-sized individuals to the occasional forest giants. This pattern of size (and hence age) structure in a forest is an indication of old-growth. Foresters and ecologists call this kind of forest all-aged (or uneven-aged), in contrast to even-aged forests that have recently developed following a catastrophic disturbance (such as severe fire) or after agricultural land is abandoned. Because trees here vary greatly in size, the canopy itself is highly stratified, comprised of trees of all heights. Viewed from above, this forest would have a "lumpy" canopy surface, unlike the relatively flat and uniform surface you would see in an even-aged plantation or young forest. You have probably also noticed that there is evidence of tree death all around; the more recently dead trees remain standing as "snags," and the trees that died longer ago have fallen to the ground and are in various stages of decay. You should be able to determine that this forest has all five key indicators of old-growth condition: (1) occasional very old (and in our case very large) trees, (2) an uneven age structure, (3) a height-stratified canopy, (4) high species diversity, and (5) abundant evidence of tree death in the forms of snags and downed logs.

Old-growth forests are important beyond their obvious beauty. They preserve a diversity of plant, animal, and microbial species. They preserve genetic diversity. They provide important habitat for some species, such as migratory songbirds that favor mature, interior forests. Finally, they provide important environmental benefits such as clean air and water. Here trees go through their entire cycle of birth, life, death, and decay, providing important functions to the surrounding forest community at all points in their lives.

A living tree stores carbon and inorganic nutrients as it grows, shading the forest floor and protecting the soil from erosion. Its leaves, flowers, and fruits are important foods for various animal species. When the tree dies, it may stand for many years, its slowly decaying wood providing food for wood-boring insects that are excavated by woodpeckers. The woodpeckers, in turn, create nesting hollows for their own use, which then become condominiums for other animal species, such as flying squirrels (*Glaucomys* spp.). Larger hollows serve as dens for mammals, including black bears (*Ursus americanus*). Once the standing dead tree topples to the forest floor, increased moisture hastens its decay, and it continues to feed innumerable fungi and bacteria while acting as a substrate for lichens and mosses. It may even serve as a "nurse log," important to the successful germination and establishment of certain tree species, such as eastern hemlock. It has been said that a tree has two lives, the first as a vigorously growing and living being and the second (of equal length) as a dead and decaying entity, slowing giving back to the forest the resources that it sequestered over many centuries.

Continue your hike, now on the Poplar Cove Loop, choosing either a clockwise or a counterclockwise circuit. This 0.75-mi loop is aptly named; although you will continue to see many species of trees (and perhaps add them to your growing list), the predominance of huge tulip-trees (*Liriodendron tulipifera*) is noteworthy. They occur singly and in small clusters all along this section of the trail, some reaching more than 20 ft in circumference, or roughly 6 ft in diameter. If you are hiking with companions, join hands to stretch your arms around the base of one of the largest tulip-trees to appreciate its enormity.

You may notice that there are few seedling, sapling, or mid-sized tulip-trees in this part of the forest. This challenges our previous conclusion that this old-growth forest is uneven-aged. Why are there no young tulip-trees, and what might their absence tell us about the history of this forest? Tulip-tree is a light-demanding, fast-growing, yet long-lived species that cannot establish in deep shade. Only large openings in the canopy provide enough sunlight for tulip-trees to grow. The current crop of ancient tulip-trees, well over 400 years old, suggests that even this forest may have established itself following a catastrophic natural disaster, such as a major windstorm or severe fire. Therefore, at least from the perspective of tulip-tree, what many ecologists consider to be one of the prime examples of old-growth forest in eastern North America may once have been even-aged. Such are the paradoxes of forest history.

As you continue your hike around the Poplar Cove Loop, you may notice something unusual about the trunks of the biggest trees: the distinctive bark patterns of various species become increasingly indistinct as the trees become very large. While you might be confident in distinguishing a middle-aged tulip-tree and a middle-aged yellow birch (*Betula alleghaniensis*) by their bark, the nondescript platey barks of their elders look surprisingly similar. You will find yourself spending increasing amounts of time craning your neck to check canopy leaves (perhaps aided by a pair of binoculars) to be sure of your tree identification.

Eventually this becomes tiring, so take note of the rich understory comprised of tree seedlings, scattered shrubs, and myriad species of understory herbaceous plants. In the spring, particularly during April and May, the herbs and shrubs steal the show with their profuse and colorful flowers. Listing every species here would be overwhelming. We'll simply note that we have seen here all five herbaceous species considered by some ecologists to be diagnostic of southern Appala-

The larva of a pipevine swallowtail butterfly feeds on a leaf of its host plant, the Dutchman's pipe. High diversity of tree species in Joyce Kilmer Memorial Forest provides a variety of habitat niches and thus supports high faunal diversity.

chian rich cove forests: sweet cicely (*Osmorhiza claytonii*); maidenhair fern (*Adiantum pedatum*); blue cohosh (*Caulophyllum thalictroides*); ginseng (*Panax quinquefolius*); and stinging nettle (*Laportea canadensis*). If you are curious about how ecologists interpret the stunning diversity of forests like this one, you may enjoy reading the sidebar "Wildflower Diversity in Temperate Forests" in hike 6 (Station Cove Falls).

When you return to the memorial plaque, take a break and consider your lengthy list of tree species. The last time we hiked here, we encountered 30 different tree species in a 2-hour circuit of the loop trails. Magnolias, tulip-trees, hemlocks, ash, beech, maples, hickories, oaks, dogwood, hemlock, birch, silverbell, basswood, and many others populate this rich cove forest. If you have visited other forests in the southern Appalachians, you will recognize that the diversity of trees here is quite impressive. Indeed, rich cove forests in the southern mountains harbor more tree *and* herbaceous species than any other temperate forest type in North America.

Continue your hike, completing the second half of the lower Memo-

Eastern North American–Eastern Asian Connections

Why do rich cove forests harbor so many species in a relatively small area? Certainly environment plays a role. Soils are deep, moist, and nutrient-rich, meeting the needs of a wide range of species. Summers are warm and mild, and the winters are not as harsh as they are at higher elevations or higher latitudes. The topography of the cove itself protects plants from the drying effects of high winds and excessive sun exposure found on the upper slopes, ridges, and summits above the cove. Further exploration of the processes maintaining high herbaceous diversity is to be found in the sidebar "Wildflower Diversity in Temperate Forests," in hike 6 (Station Cove Falls). However, a more fruitful question at this point might be "Where did all these species come from?" This is a question relevant not only for rich coves but for all the natural communities of the southern Appalachians.

The answer takes us back to the later part of the Mesozoic Era, the Age of Reptiles, when dinosaurs roamed the earth. The ancestors of the tree species you see today were already evolving and diversifying in subtropical and warm-temperate climatic regions that were widespread at the time. The dinosaurs became extinct at the end of the Mesozoic (65 million years before the present), but subtropical and warm-temperate forests flourished throughout much of the next great (and present) era, the Cenozoic, the Age of Mammals.

Early to mid-Cenozoic fossil records from across much of North America, Europe, and Asia indicate that such forests were widespread for many millions of years, and they harbored the ancestors of the tree species we see today in rich cove forests of the southern Appalachians. But later in the Cenozoic, climates cooled worldwide, culminating in the Ice Age (Pleistocene Epoch) of the past 2 million years. Many parts of the subtropical and temperate zones also became drier, leaving fewer places where species-rich, warm-temperate forests could flourish. Today, these species-rich, warm-temperate forests are found primarily in eastern North America, in eastern Asia, and to a lesser extent, in western North America and western Europe.

You might ask if some of the many plant families and genera shared between eastern North America and eastern Asia might have gotten around by various means after they evolved. Indeed, North America and eastern Asia have been connected from time to time by land bridges that would have facilitated overland dispersal via gradual migration. Closely related species recently isolated in eastern Asia and eastern North America (after the land bridges were no longer available) may have traversed oceanic barriers in events of long-distance dispersal. Scientists are still trying to understand the roles play by migrations and long-distance dispersal in the fascinating similarities we see today between these two distant regions.

Today, the best examples of similar forests are found in protected coves in the southern mountains of eastern North America and eastern Asia, particularly in China and Japan. If you could be magically transported from the southern Appalachians to one of those Asian forests, you would feel right at home, readily able to identify many of the trees to their proper family and/or genus. Your Asian colleague would feel equally at home in the southern Appalachians.

rial Loop on your way back to the trailhead and parking area. During this final part of the hike, stop and take one last look at the cathedral-like stand with its towering trees. It is tempting at this point to think of the forest here as ancient, unchanging, frozen in time. But even the longest-lived of eastern trees, like the mighty tulip-trees and hemlocks, must eventually die. Careful ecological studies of old-growth forests, including Joyce Kilmer Memorial Forest, have revealed much about their dynamics. One study conducted here found that 5.8% of canopy trees died during every decade, on average. Another study revealed that gaps created by falling trees represented 17.3% of the canopy at any time. Such openings allow sunlight to reach the forest floor, stimulating seedling and sapling trees to begin their own ascent to the canopy. As old gaps are filled by young and growing trees, new ones are created by the death of forest giants. This explains the paradox of how the even-aged forest of tulip-trees, which germinated long ago, eventually became all-aged and considered old-growth.

The cycle of regeneration, growth, maturity, death, and more regeneration makes this a very dynamic forest. Unless you are fortunate enough to witness the falling of a massive canopy tree (from a safe distance, of course), you will be unable today to perceive the cyclic drama that plays out over hundreds of years. However, you can gather evidence of forest dynamics from observations of forest age structure, the distribution of dead and downed trees, and the gaps they create.

A sign at the entrance kiosk shares "Trees" and commemorates World War I veteran and poet Joyce Kilmer. Enormous tulip-trees along the trail add to the forest's complex structure and give the forest a primeval feeling. Long after their death, fallen trees provide habitat for the forest floor community, restarting the circle of life.

As you finish your hike and return to the parking area, you might speculate about how a new cycle of mortality, caused by the devastating effects of hemlock woolly adelgid on eastern hemlock, will affect this forest in the future. In even the stateliest of old-growth forests, change is inevitable.

15. GREGORY BALD

Why Are We Here?

This long and often strenuous hike will take you through a complex landscape supporting numerous natural communities. As you climb and move through different topographic positions across a range of elevations, the character of the forest will change, sometimes gradually, sometimes more abruptly. You will thus experience a continuum of environmental conditions supporting a parallel continuum of natural communities. The effects of past and present human influences will also be evident to a greater or lesser extent. This idea may seem rather abstract, but you will be amply rewarded for your time and effort with a big-picture perspective of how environmental factors and human history dictate the complex of natural communities spread across the southern Appalachian landscape.

During this hike, you will encounter, in sequence, cove forest, oak forest, montane pine forest and woodland, montane oak forest, northern hardwood forest, and grass bald. In each of these communities, you will learn about the environmental conditions and human interventions (both positive and negative) responsible for their present-day composition and structure.

Hikers reaching Gregory Bald will also be rewarded with spectacular views of the western Great Smoky Mountains. An array of seasonal highlights on Gregory Bald includes the vibrant display of flowering flame azaleas (*Rhododendron calendulaceum*) in June, blueberries in summer, and colorful foliage in fall.

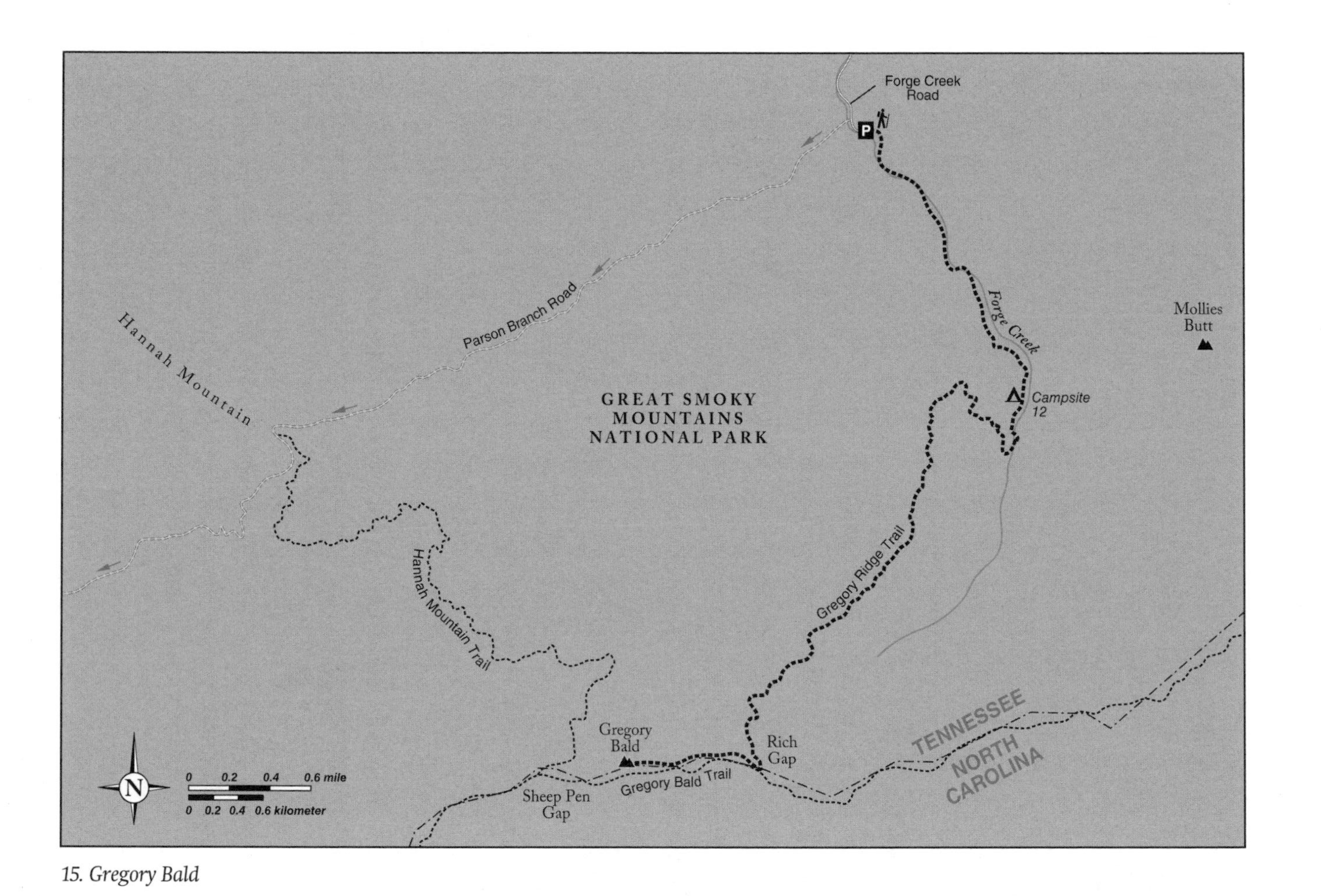

15. *Gregory Bald*

15. GREGORY BALD

Highlights: An ecological journey through the forest primeval, traversing old-growth forests as you climb toward Gregory Bald. Gregory Bald is a worthwhile destination in its own right, offering spectacular views of the western Great Smoky Mountains.

Natural Communities: Cove forest, oak forest, montane pine forest and woodland, montane oak forest, northern hardwood forest, grass bald

Elevation: 2,000 ft (trailhead); 4,949 ft (Gregory Bald summit)

Distance: 5.0 mi (one-way) from trailhead to Rich Gap on the Gregory Ridge Trail; additional 0.7 mi (one-way) from Rich Gap to Gregory Bald on the Gregory Bald Trail. Round-trip hike from the trailhead of the Gregory Ridge Trail to the summit of Gregory Bald is thus 11.4 mi.

Difficulty: Strenuous hike to Rich Gap on the main crest of the Smokies and thence to Gregory Bald. But you can turn around anytime to hike back.

Directions: To reach the trailhead of the Gregory Ridge Trail, drive the first half of the one-way Cades Cove Loop Rd. to the Cade's Cove Visitor Center (6.4 mi from entrance to loop road), then drive an additional 1.4 mi on the two-way Forge Creek Rd., which ends at the trailhead (no facilities) and a small parking area shortly after you pass the turnoff (to the right) for Parson's Branch Rd. (one-way, out of the park). On your departure, return to the Cades Cove Visitor Center on the Forge Creek Rd., then complete the remaining half of the Cades Cove Loop Rd.

GPS: Lat. 35° 33′ 45.7″ N; Long. 83° 50′ 45.2″ W (trailhead)

Information: Great Smoky Mountains National Park, 107 Park Headquarters Rd., Gatlinburg, TN 37738, (865) 436-1200

The Hike

You'll begin this hike by climbing and then immediately descending a rocky ridge. After you descend, you'll be hiking through young, second-growth forest; this forest continues after you cross a log footbridge on Forge Creek.

The trail now follows an old road through the even-aged, second-growth forest. Common trees in this forest are white pine (*Pinus strobus*), eastern hemlock (*Tsuga canadensis*), red maple (*Acer rubrum*), tuliptree (*Liriodendron tulipifera*), black locust (*Robinia pseudoacacia*), Fraser magnolia (*Magnolia fraseri*), and pitch pine (*Pinus rigida*). The understory is open, except for a dense shrub layer of rosebay rhododendron (*Rhododendron maximum*) and doghobble (*Leucothoe fontanesiana*) on the banks of Forge Creek.

The paucity of herbaceous species in the understory and canopy suggests that the soil here is acidic and nutrient-poor. According to the Great Smoky Mountains Association's *Hiking Trails of the Smokies*, the forest along this section of Forge Creek was cut to make charcoal that fueled a small iron foundry once located along the creek. After familiarizing yourself with the structure and species composition of this young forest, watch carefully for larger trees as you continue your gentle ascent along Forge Creek.

At about 0.7 mi, the old road ends, the trail narrows, and the forest begins to accumulate additional canopy species typical of the rich cove forest that lies ahead. At 1.0 mi, a careful observer will note a definite shift in the forest's age structure. Towering above the smaller trees are larger individuals of tulip-tree and eastern hemlock, all about 3 ft in diameter. These large trees signal an important change in forest structure and management history; their greater size and age suggest that this forest, if ever logged, was cut selectively and much less recently than the forest you have traversed thus far. Unfortunately, at the time of our last visit, the hemlocks on this hike had already succumbed to the hemlock woolly adelgid (*Adelges tsugae*). The future composition of this forest thus lies in doubt because hemlock was found nearly throughout and dominated the forest canopy, especially near the creek.

As you continue hiking along the creek, you will soon encounter large specimens of many tree species, indicating your passage into the forest primeval, a rich, old-growth cove forest. Those so inclined may wish to begin a species list of the canopy trees along this stretch of Gregory Ridge Trail, for much of the second mile of the trail passes through this rich cove forest community. With a good tree guide and not too much effort, you can easily compile a list of 25 or more tree species, including those seen earlier in the second-growth forest, plus many others, such as basswood (*Tilia americana*), cucumber magnolia (*Magnolia acuminata*), American beech (*Fagus grandifolia*), and silverbell (*Halesia tetraptera*).

When your neck tires of looking up at the canopy, turn your attention to the forest floor. As the canopy becomes more diverse, so does the understory, and the second mile of the trail features a rich collection of understory plants, including many of the spring wildflowers for which these rich cove forest communities are well known.

The steep-sided valley of Forge Creek restricts the development of the rich cove forest to some extent. Look toward the creek and note that the valley bottom remains a thicket of rosebay rhododendron and doghobble under the eastern-hemlock-dominated canopy. The steep slopes above the valley bottom support oak forest. The rich cove forest is thus best developed in bands just above, and parallel to, Forge Creek. Fortunately, this stretch of the trail ascends through one of these bands on the northeast-facing slope above Forge Creek.

A second footbridge crosses Forge Creek at 1.8 mi, and soon thereafter a third footbridge crosses again. The trail then begins its ascent of Gregory Ridge but soon passes wilderness campsite #12 at 2.0 mi. You may wish to stop here for a rest, enjoying the last sound of the rushing waters of Forge Creek before the 3-mi ascent of Gregory Ridge to Rich Gap. Before you get out a well-earned snack, look around to be sure that there are no black bears in the vicinity. Our last visit to this site included a surprise encounter with a curious young black bear!

When you depart campsite #12, you'll climb steeply onto the flank of Gregory Ridge. Watch carefully for the first large chestnut oaks (*Quercus montana*), with their scalloped leaf edges, evidence that you are departing the rich cove forest and briefly entering an oak forest. As the trail climbs, it will alternately flank small, drier ridges, then cross more protected ravines. Note how these changes in topographic position affect community makeup. The ridges support the drier oak forest, with its understory of mountain laurel (*Kalmia latifolia*) and bear huckleberry (*Gaylussacia ursina*), both in the heath or blueberry family. The ravines, cooler and moister because of their protected topography, support canopy species more characteristic of the cove forest below, including basswood, tulip-tree, silverbell, and eastern hemlock. Even the species-rich understory briefly reappears on the sides of these ravines, while rosebay rhododendron fills the ravine bottoms.

At about 2.5 mi, the canopy begins to open, and the trailside understory vegetation becomes increasing dominated by low-growing evergreen species, including round-leaved galax (*Galax urceolata*), trailing arbutus (*Epigaea repens*), and aromatic wintergreen (*Gaultheria procumbens*). You are climbing up the terminus (locals call it a "lead") of an exposed, steeply sloping ridge; shallow soils and exposure to intense sunlight and drying effects of winds make this a habitat where drought-tolerant trees like pines can thrive.

Here you *may* find a montane pine forest and woodland dominated by pitch and Table Mountain (*Pinus pungens*) pines. We say *may* because hikers will immediately notice that most of the large pines on this site were killed by a prescribed fire that was set by the US Park Service in 2007. Why would the agency responsible for stewardship of this forest intentionally burn it, knowing that the fire would kill the canopy pines?

In fact, pine stands on steep, rocky slopes like this one require periodic fire to remain healthy. Hot fires passing through a mature pine stand remove the aging canopy trees as well as much of the "duff" layer, composed of spongy, partially decomposed organic matter that slowly accumulates on the forest floor. This creates conditions that favor pine regeneration. Without fire, the duff layer thickens, establishing a barrier that prevents the roots of pine seedlings from successfully reaching the mineral soil beneath. Instead, species of drought-tolerant hardwoods successfully colonize the site, creating dense shade that also interferes with the light-demanding (or shade-intolerant) pine seedlings. Both pitch pine and Table Mountain pine are specially adapted for a fire-prone habitat, and their different adaptations are discussed in the sidebar "A Tale of Two Pines" in hike 22 (Table Rock).

Without fire, a pine stand will become increasingly dominated by hardwoods such as scarlet oak (*Quercus coccinea*), black gum (*Nyssa sylvatica*), sourwood (*Oxydendrum arboreum*), red maple, and sassafras (*Sassafras albidum*) and may remain that way indefinitely. In the past, fires set by both humans and lightning strikes resulted in ideal conditions for renewal of pine forests. We know that this stand experienced three major fires in the century preceding 1940. However, since 1940, a US Park Service policy of fire suppression prevented any fire at this site until the prescribed fire in 2007.

The 2007 fire was possible because of a change in Park Service policy in 1996 allowing lightning-ignited fires to burn freely in areas where they present no risk to human life or property. Since establishment of this "let burn" policy, the Park Service has managed several lightning-ignited fires in Great Smoky Mountains National Park; most were small in size, the largest covering about 1,000 acres. When you reach the prominent sandstone outcrop at a sharp bend in the trail, pause to consider the current status of this stand. What is the mix of

Rob Klein, the fire crew leader for Great Smoky Mountains National Park, demonstrates the depth of the duff layer at Gregory Ridge, despite the 2007 prescribed burn. Additional burns would be needed to remove the duff and restore the pine stand.

deciduous and coniferous species? Can you find any evidence of establishment of pine seedlings after the fire in 2007? Can this stand be restored to a montane pine and woodland community?

The National Park Service had planned prescribed fires in 5-year intervals for the next 20 years, with the goal of further reducing the accumulated duff layer and restoring the stand to a montane pine forest and woodland. However, this plan is too costly and logistically

A few canopy pines remain at Gregory Ridge, but clearly the stand is being taken over by a mix of drought-tolerant hardwoods and evergreen shrubs in the Ericaceae family. Pines cannot regenerate under a dense understory like the one now in place.

difficult, so this pine stand will have to wait for the next well-placed lightning strike. Pine stands such as these are slowly disappearing all over the southern Appalachians because of fire suppression policies. At this point, the chances of success using prescribed fire are too low to make this Gregory Ridge pine stand a top priority in a large park with countless land management needs. For more on the significance of fire in southern Appalachian forests, please refer to the sidebar "Fire on the Mountain" in hike 21 (Babel Tower and Wiseman's View).

After resting at the sandstone outcrop, resume your hike, shifting to the flank of Gregory Ridge. You will continue through the montane pine forest and woodland for a while, but gradually the montane oak forest reasserts itself as you leave the drier part of the ridge. Dominance of chestnut oak is short-lived, however, as you continue hiking. Soon the northwest-facing slope below the crest of Gregory Ridge is once again dominated by canopy and understory species that you saw earlier in the rich cove forest during the first part of the hike.

Along this stretch of the trail, soils are deep, and the northwest aspect and increasingly higher elevation contribute to cooler and moister conditions. The soils appear fertile, and you soon find yourself walking through a lush, old-growth forest dominated by species such as basswood, silverbell, maples, tulip-tree, black locust, and magnolias. Northern red oak (*Quercus rubra*) begins to assume greater importance in this stand, but chestnut oak is still present. The understory is similarly rich, and there is a diverse herbaceous layer beneath scattered shrubs.

At about 3 mi, the trail runs briefly along the ridgetop, with an associated shift toward more drought-tolerant species. Soon the trail resumes its course along the northwest-facing flank of the ridge, and the old-growth forest here is one of the most beautiful in Great Smoky Mountains National Park. Red oak is abundant, but dominance is shared by numerous tree species, including white ash (*Fraxinus americana*), yellow buckeye (*Aesculus flava*), black cherry (*Prunus serotina*), cucumber magnolia, sweet birch (*Betula lenta*), and sugar maple (*Acer saccharum*).

The understory here is predominantly herbaceous, and thus the forest is open, affording sweeping views through the scattered large trees. Placing this magnificent forest in a single, simple category is difficult because there are intermingled elements of rich cove forest, montane oak forest, and northern hardwood forest. The result is a beautiful old-growth forest to enjoy along the rest of your hike to Rich Gap.

At the 5-mi mark, you'll reach the crest of the Great Smokies at Rich Gap where the trail intersects the Gregory Bald Trail at 4,500 ft. Catch your breath, then turn right to walk west along the level ridgeline trail toward Gregory Bald, just over a half mile away. You'll notice that the canopy height is much shorter than it was on the upper section of the Gregory Ridge Trail. This forest is probably just as old as the one below, but soils are thinner here and the ridgeline is more exposed, limiting the growth of trees here. Notice that lower-elevation dominants like tulip-tree have disappeared and the forest has morphed into a northern hardwood forest, dominated by yellow birch (*Betula alleghaniensis*), sugar maple, northern red oak, and American beech.

As you continue the ridgeline trail out to Gregory Bald, you may notice pockets of a special subtype of northern hardwood forest called a beech gap, to your right, where beech grows in gnarled, nearly pure stands. This northern hardwood subtype can be found in the "gaps," a local term for low points along the ridges. These more protected areas along the ridgeline also harbor more herbaceous species, such as sedges and spring wildflowers. During our most recent visit, we noticed considerable damage from the European wild boar (*Sus scrofa*) along this ridgeline trail. Look for areas where the soil appears to be tilled; the boars use their snouts to dig for roots, and this activity has had a considerable impact on some spring wildflower species. European wild boar were introduced on Hooper Bald, another of our hikes, and there is additional information there about this obnoxious pest. In addition to poaching by pigs, beech gaps have been decimated by beech bark disease, which is the association of a beech scale insect (*Cryptococcus fagisuga*) and a fungus (*Nectria coccinea* var. *faginata*), so if you spot these stands of beech, look for die-back.

At 0.6 mi, you'll see the forest grading into shrubs in front of you, and you'll emerge out of the forest onto Gregory Bald. If you are here in June, you will be dazzled, as we were, by the scorching colors of flame azalea. William Bartram, who first described flame azalea in his *Travels* (1791), wrote, "The epithet fiery I annex to this most celebrated species of azalea, as being expressive of the appearance of its flowers; which are in general of the color of red-lead, orange, and bright gold, as well as yellow and cream-color."

As your eyes adjust to the brightness of the open sky and the colorful array of flame azaleas, you'll notice that the trail continues along the bald in a due west direction. Although we classify this natural community as a grass bald, only human intervention in the form of mowing keeps back encroaching flame azalea and highbush blueberry (*Vaccinium corymbosum*). Indeed, historical photographs show that Gregory Bald was much more open in the past, likely due to early settlers bringing livestock up to graze the abundant grasses during the summer months. With the cessation of grazing, shrubs and trees naturally encroach in these open spaces. Further reading on grass balds may be found in the sidebar "Grass Bald Origins and Future" in hike 13 (Hooper Bald).

Continue to follow the trail through the center of the bald until

Community Concepts in Conflict

With so many different natural communities (grass bald, northern hardwood forest, beech gap, montane oak forest, montane pine forest and woodland, oak forest, and cove forest) found along the Gregory Ridge Trail, this hike illustrates beautifully how communities intergrade and overlap. Determining where one natural community ends and another begins is a difficult task, because each community is defined by a suite of individual trees and other plants, each with its own environmental needs and life history strategies. As a result, most communities are present along a continuum of the ranges of overlapping individual species.

Some boundaries are distinct; for example, note the sharp boundary between the grass bald of Gregory Bald and the adjacent montane oak forest. Part of this visual boundary is structural (trees versus grasses), but if you look carefully, you'll also notice that there are few species shared between the two communities. On the other hand, the boundary between the northern hardwood forest and the montane oak forest is far less distinct; at least two of the most important species, northern red oak (*Quercus rubra*) and yellow birch (*Betula alleghaniensis*), are dominant in both communities. Here it's impossible to draw a boundary line. You'll find this pattern of both distinct and indistinct boundaries repeated throughout the southern Appalachians and, indeed, in many other natural landscapes.

The existence of both distinct and indistinct boundaries between natural communities in landscapes helped fuel one of the most interesting, and sometimes acrimonious, debates in the history of plant ecology. In the early 20th century, the prevailing view was that natural communities were readily distinguishable at sharp boundaries or "ecotones" where different natural communities adjoined. The distinctness of such real and organic entities was believed to be maintained by specific environmental requirements of a group of tightly coadapted and interacting species.

The idea that natural communities were well-defined associations of species having very similar environmental preferences was stimulated by the extensive published research and influential personality of Frederic Clements at the University of Nebraska. He believed that natural communities were both predictable and tightly interactive, behaving almost like superorganisms. In Clements's view, a particular natural community would occupy most of a given landscape under a given climate.

In opposition to Clements's ideas were those of Henry Gleason, originally from Illinois. Gleason proposed that natural communities were continuously varying assemblages of species, each reflecting the distinctive and individualistic preferences of species for various combinations of environmental conditions. In Gleason's view, an observer moving along a gradient of environmental conditions (such as those associated with changing elevation in mountainous terrain) would experience varying mixtures of species as the component species waxed and waned in their abundances, rather than distinctive, tightly integrated natural communities.

Although Clements's ideas held sway during the first half of the 20th century, particularly in the United States and Great Britain, ecologists in the latter half of the 20th and early 21st centuries have tended to side with Henry Gleason. On your hike to Gregory Bald, we ask you to consider the distributions of individual species, like chestnut (*Quercus montana*) and red (*Q. rubra*) oaks, and how these distributions contribute to the nature of the forest community. Based on your observations, do you feel more "Clementsian" or "Gleasonian"?

you reach the US Geological Survey marker, which notes the elevation as 4,949 ft. If you continue to face west, look back toward the four o'clock position (north-northeast), and you'll see the open areas of Cades Cove below you. This is the perfect place for a picnic, which you can augment with ripe blueberries if you happen to be here in August. On a sunny day, this is a lovely place to linger.

On your return journey, focus your attention on the boundaries between communities. Where does one natural community end and another begin? For more discussion of this topic, please refer to this hike's sidebar.

Notice as you turn left to head down the Gregory Ridge Trail from Rich Gap that you're gaining species as you descend. We counted one-third of the number of species on the ridgeline trail going to Gregory Bald compared with this upper portion of the Gregory Ridge Trail. As you descend, look for large specimens of black cherry and Carolina silverbell, which were absent from the ridgeline trail going out to Gregory Bald. Farther down, notice that tulip-tree, dominant in the first section of old-growth forest, is still absent. As you lose elevation, look for the appearance of the four-lobed leaves and straight trunks of this species—it reappears at about 4,200 ft. Choose one or two other species you can readily identify and note when they first appear, or disappear, from the forest along the trail.

Enjoy your return trip (which is mostly downhill) and challenge yourself to see both the forest *and* the trees. Concepts of natural communities are, for the most part, a human construction. We look for patterns that help us make sense of the beautiful complexity of nature that is illustrated so clearly on this journey.

---

16. CARLOS CAMPBELL OVERLOOK

Why Are We Here?

The Carlos Campbell Overlook is widely recognized as an exceptional place to observe, in one vista, most of the natural communities of the Smokies, arrayed in predictable positions relative to climatic gradients dictated by topography. Ecologists familiar with the Smokies call this stop the Whittaker Overlook because of its association with Robert H. Whittaker, author of the classic 1956 paper "Vegetation of Great Smoky Mountains." It was here that Whittaker established a model that uses environmental factors, primarily elevation and topographic moisture, to predict the distribution of natural communities in the Smokies. The Whittaker Overlook, as we shall refer to it, also affords an opportunity to observe the devastating effects of several introduced insect pests and plant diseases on the natural communities of the region.

The Hike

Not a hike but a roadside stop, the Whittaker Overlook is nevertheless a must-see opportunity for any hiker traveling in the region. The overlook faces east toward Balsam Point, a 5,840-ft summit on the shoulder

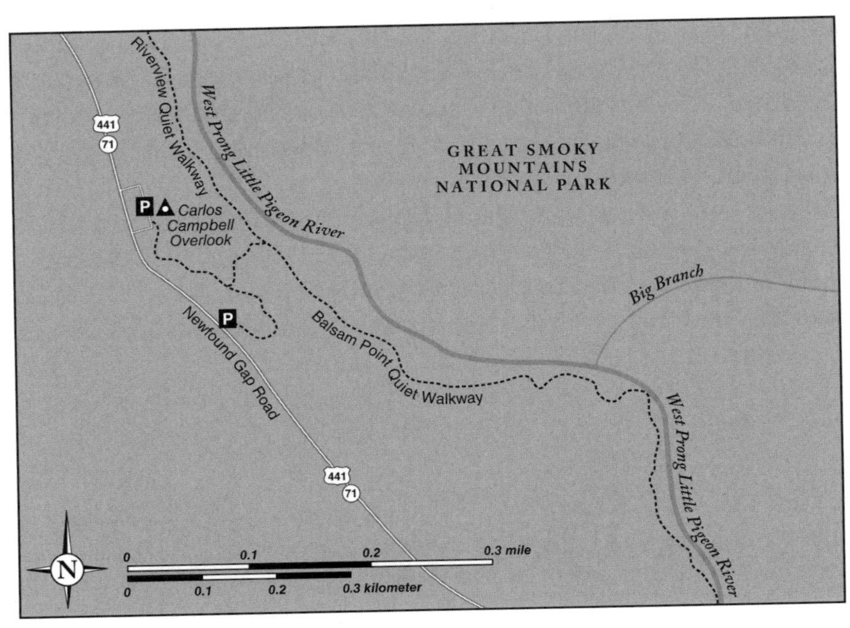

16. Carlos Campbell Overlook

of Mount LeConte. Here you'll see a deep valley flanked by two parallel ridges that lead your gaze upward toward Balsam Point. This is truly a place where one can see the forest with the trees. Binoculars will help you see the distinctions more readily.

As you take in the panoramic view, you can see the communities described so eloquently by Robert Whittaker. Study the expanse of green until you can discern differences in the colors and textures of the forests carpeting the mountains. The deep green spruce-fir forests occupy the highest summits, confined in your view here to Balsam Point. Although the summit is still forested (mainly by red spruce [*Picea rubens*]), the forest is studded with the white ghosts of Fraser firs (*Abies fraseri*) that have succumbed to the onslaught of the balsam woolly adelgid (*Adelges piceae*), a tiny, bark-feeding insect introduced from Europe.

Below the Balsam Point summit, in the contours of the mountain's deep ravines, look for the gray skeletons of dead eastern hemlock forests (*Tsuga canadensis*), where they were protected from the harsher conditions at higher elevations in cove forests. Most of the mighty

16. CARLOS CAMPBELL OVERLOOK

Highlights: Scenic vista featuring most of the common natural communities in the southern Appalachian Mountains

Natural Communities: Spruce-fir forest, northern hardwood forest, shrub bald, cove forest, montane pine forest and woodland, oak forest

Elevation: 2,000 ft

Distance: N/A

Difficulty: N/A

Directions: The Carlos Campbell Overlook is located on US 441 on the Tennessee side of Great Smoky Mountains National Park between Newfound Gap and the Sugarlands Visitor Center. From the Newfound Gap parking area, drive 11 mi. Look for the pullout on the right, which is the Carlos Campbell Overlook. The view is slightly better from the second pullout, just past the Carlos Campbell overlook (no facilities).

GPS: Lat. 35° 39′ 29.6″ N; Long. 83° 31′ 12.5″ W (overlook)

Information: Great Smoky Mountains National Park, 107 Park Headquarters Rd., Gatlinburg, TN 37738, (865) 436-1200

hemlocks are dying or already dead, the result of an infestation by another exotic invasive insect, the tiny hemlock woolly adelgid (*Adelges tsugae*).

Various high-elevation mixtures of northern hardwoods, grading into both acidic and rich cove forests, occupy the broader and more gently sloping ravines above and below the hemlock. One of the more prominent features you'll note in any season is the ridge to the left of Balsam Point that has a flattened appearance to the vegetation; this is a large shrub bald on a dry, south-facing ridge, comprised of members of the blueberry family such as mountain laurel (*Kalmia latifolia*) and Catawba rhododendron (*Rhododendron catawbiense*).

The ridges nearer you, on either side of Balsam Point, are about a thousand feet below the Balsam Point summit, about 4,250 ft. You will see extensive die-off here as well, but these corpses are pines, the remnants of a montane pine forest and woodland that once thrived on the dry, rocky, exposed ridges. These habitats are susceptible to lightning strikes and thus frequent fire. Today these pine forests are filled with the bleached but still standing dead stems of the dominant trees, pitch pine (*Pinus rigida*) and Table Mountain pine (*P. pungens*). A

The Whittaker Overlook has in one view most of the natural communities found in Great Smoky Mountains National Park. The center peak, which is much taller and farther away, is Balsam Point. Look for the deep green of spruce-fir forest there (1). Down the flanks of the mountains, you can see mortality of eastern hemlock trees in the ravines, interspersed with northern hardwood forests (2). The flattened vegetation to the left of Balsam Point is a shrub bald (3). Mortality of pines can easily be seen on the two ridges nearest you (4); these stands are montane pine forests and woodlands. Below is the backdrop of oak forests on the lower slopes (5). Finally, the even canopy in the foreground (6) is comprised of second-growth tulip-trees that germinated following clearing.

combination of long-term fire suppression policy, drought, and the southern pine beetle (*Dendroctonus frontalis*) has jeopardized these forests all over the southern Appalachians.

The lower slopes below these ridges contain a variety of oak forests, the default backdrop forest in the southern Appalachians. Finally, in the foreground, you can see where the oak-dominated forests give way to a younger, even-aged forest dominated by tulip-tree (*Liriodendron tulipifera*), a legacy of an earlier time when arable land of lower elevations was cleared and farmed by settlers. Depending on the season, you might notice the smooth appearance of the even-aged canopy, the slightly pointed crowns, or the characteristic lime green color of tulip-trees in early spring; in September, you might instead notice the characteristic golden hues that signify early fall color. Over time, the moister parts of this second-growth forest will continue to be dominated by tulip-tree and become a rich cove forest, while drier areas may support an oak forest typical of lower elevations in the southern Appalachians.

The devastation of three of the most significant forested ecosystems of the southern Appalachians is in plain view. Consider the role that humans have played in bringing our mighty southern Appalachian forests to such a sad juncture. In the case of the Fraser fir, the balsam woolly adelgid was accidentally introduced from Europe and reached the southern Appalachians in the 1950s. The effect of the balsam woolly adelgid on our endemic firs may have been exacerbated by stresses on the trees brought about by a variety of anthropogenic effects on air quality, including acid rain and high ozone levels. In the case of eastern hemlock, the hemlock woolly adelgid is an exotic insect pest from Asia, introduced to the United States in the 1920s and reaching the Smokies in 2002. You may learn more about this tragic situation in the "Hemlock Woolly Adelgid" sidebar in hike 18 (Ramsey Cascades). The cause of mortality of the pines is a complex interaction of drought, outbreaks of southern pine beetle, and fire suppression. While the first two of these factors are natural, the third is distinctly anthropogenic. More about the important role of fire in maintaining southern Appalachian pine forests may be found in the "Fire on the Mountain" sidebar in hike 21 (Babel Tower and Wiseman's View).

These catastrophic forest declines visible from the Whittaker Overlook can be seen throughout the southern Appalachians. Equally troubling are other changes not immediately visible from the Whittaker Overlook. Whittaker himself witnessed the devastation caused by the chestnut blight (*Cryphonectria parasitica*), which had already decimated the American chestnut (*Castanea dentata*) when he was doing his inventory work in the Smokies in the late 1940s. We discuss the chestnut blight in greater depth in the "Where Has the Chestnut Gone?" sidebar in hike 27 (Mount Jefferson). American beech (*Fagus grandifolia*) is succumbing to beech bark disease (*Nectria* spp.), flowering dogwood (*Cornus florida*) has been decimated by dogwood anthracnose (*Discula destructiva*), butternut (*Juglans cinerea*) has fallen prey to butternut canker disease (*Sirococcus clavigigenti-juglandacearum*), and American elm (*Ulmus americana*) is subject to Dutch elm disease (*Ophiostoma ulmi* and *O. nova-ulmi*). All five of the aforementioned diseases are caused by fungal pathogens known or believed to be of exotic origin.

Other biological threats to the integrity of southern Appalachian forests are already at our doorstep. Will gypsy moth (*Lymantria dispar*)

A Call to Action

It is easy to feel powerless as we witness the far-reaching destruction wrought by invasive exotic species, which is so evident at the Whittaker Overlook and throughout the southern Appalachians. What can we, as citizen-scientists, do to reduce the risk of further devastation of our forests by exotic pests and diseases? Three approaches to risk reduction include supporting federal programs, working through nongovernmental organizations, and contributing individual actions.

Government agencies enforce laws that protect our native flora and fauna. As one example, the US Department of Agriculture's Animal and Plant Health Inspection Service (APHIS) is charged with preventing the importation of invasive exotic plants and harmful plant pests to the United States. APHIS is currently developing new quarantine regulations for importation of plant nursery stock. When implemented, these regulations will greatly reduce the risk of additional invasive species, from the imported plants either becoming invasive themselves or serving as vectors for dangerous plant pests. APHIS is also improving its risk analyses for plants and plant pests and working to target pest problems at their points of origin, rather than where they enter the United States. The development of early warning systems and more effective responses to introduced pests will complement approaches aimed at preventing introduction of invasive plants or plant pests.

Many nongovernmental organizations, funded privately by citizens such as ourselves, prevent or ameliorate the devastating effects of introduced pests and diseases. Botanical gardens and native plant societies educate the public about threats from invasive species. Organizations such as the American Chestnut Foundation (ACF) have worked on individual species. For years, the ACF has worked to back-cross the American chestnut (*Castanea dentata*) with the blight-resistant Chinese chestnut (*C. mollissima*), with the goal of producing a blight-resistant American chestnut. The Center for Plant Conservation heads a network of botanical centers and coordinates the nationwide effort to conserve native plant material, either as live plants or seed, for future restoration efforts. Finally, these organizations partner with the many land conservancies that protect, manage, and restore natural ecosystems.

At the individual level, there are many things we can do to reduce the importation and/or spread of exotic pests. As gardeners or landscapers, we can use native and/or locally sourced plant materials for gardening and landscaping when appropriate. When working with exotic plant materials, we must ensure that these come from reputable sources known to be conforming to applicable laws and regulations related to their importation. Firewood consumers should also be sure that the wood they burn is locally sourced to prevent spread of wood-infesting pests like the emerald ash borer (*Agrilus planipennis*). Finally, we can assist government agencies by actively reporting new pests or infestations to state extension agents.

Our southern Appalachian forests are in peril from the introduction of exotic pests, but we remain hopeful for the future. Awareness of the problems posed by exotic pests and efforts to address these problems has grown. Continued efforts by governmental and private organizations, with the support of increasing numbers of citizen-scientists like ourselves, will attack the issue on multiple fronts and may bring some of our most treasured resources back from the brink of extinction.

devastate our forests as it has forests to the north? Will emerald ash borer (*Agrilus planipennis*) eliminate our mighty ashes (*Fraxinus* spp.)? As we come to fully appreciate the magnitude of the changes just described, we can conclude three things: (1) our southern Appalachian forests are undergoing profound changes, resulting from decimation of one foundation species after another; (2) over the long evolutionary and ecological history of these forests, these rapid changes in a period of decades are contributing to what scientists are calling the sixth major extinction event in earth's history; and (3) in every case mentioned, human intervention, most often through introduction of exotic pests, is the direct cause of the devastating changes now occurring. The sidebar suggests ways we should get involved in preventing additional devastating environmental effects of introduced exotic pests.

17. ALUM CAVE

Why Are We Here?

There are many excellent reasons for hiking to Alum Cave. Foremost among these, at least from the perspective of forested (and not-so-forested) ecosystems, is that this hike affords excellent access to a well-developed shrub bald (also known as a heath bald or heath slick). In studying the shrub bald below Alum Cave, we will see evidence of how the underlying geology has had profound effects on the plant community by creating chronically unstable and steep slopes with shallow, acidic soils.

On our way to the shrub bald, we'll observe how destructive flooding and landslides can shape the riparian landscape and adjacent steep slopes in mountainous terrain. We'll reflect on how earth's most fundamental geologic process and catastrophic upheavals can lead to striking and beautiful communities that one can enjoy on a pleasant day hike during almost any season of the year. The aesthetic rewards to the Alum Cave hiker are many: outstanding floral displays of mountain laurel (*Kalmia latifolia*) and Catawba rhododendron (*Rhododendron catawbiense*) in May and June are certainly attractions, as are the spectacular views down and across the valley of the West Prong of the Little Pigeon River. Bird enthusiasts will have an opportunity to spot peregrine falcons and ravens from the overlooks.

The Hike

From the trailhead, cross Alum Cave Creek on a substantial wooden bridge. Take time to look both upstream and down. You are in the upper-elevation reaches of acidic cove forest transitioning to northern hardwood forest and spruce-fir forest, with a canopy consisting primarily of yellow birch (*Betula alleghaniensis*) and scattered eastern hemlock (*Tsuga canadensis*). The predominance of yellow birch here is likely an effect of frequent catastrophic disturbance (flooding). Note also that the understory is a dense stand of rosebay rhododendron (*Rhododendron maximum*) with a ground layer of mountain doghobble (*Leucothoe fontanesiana*) below. The locals call this a "rhododendron

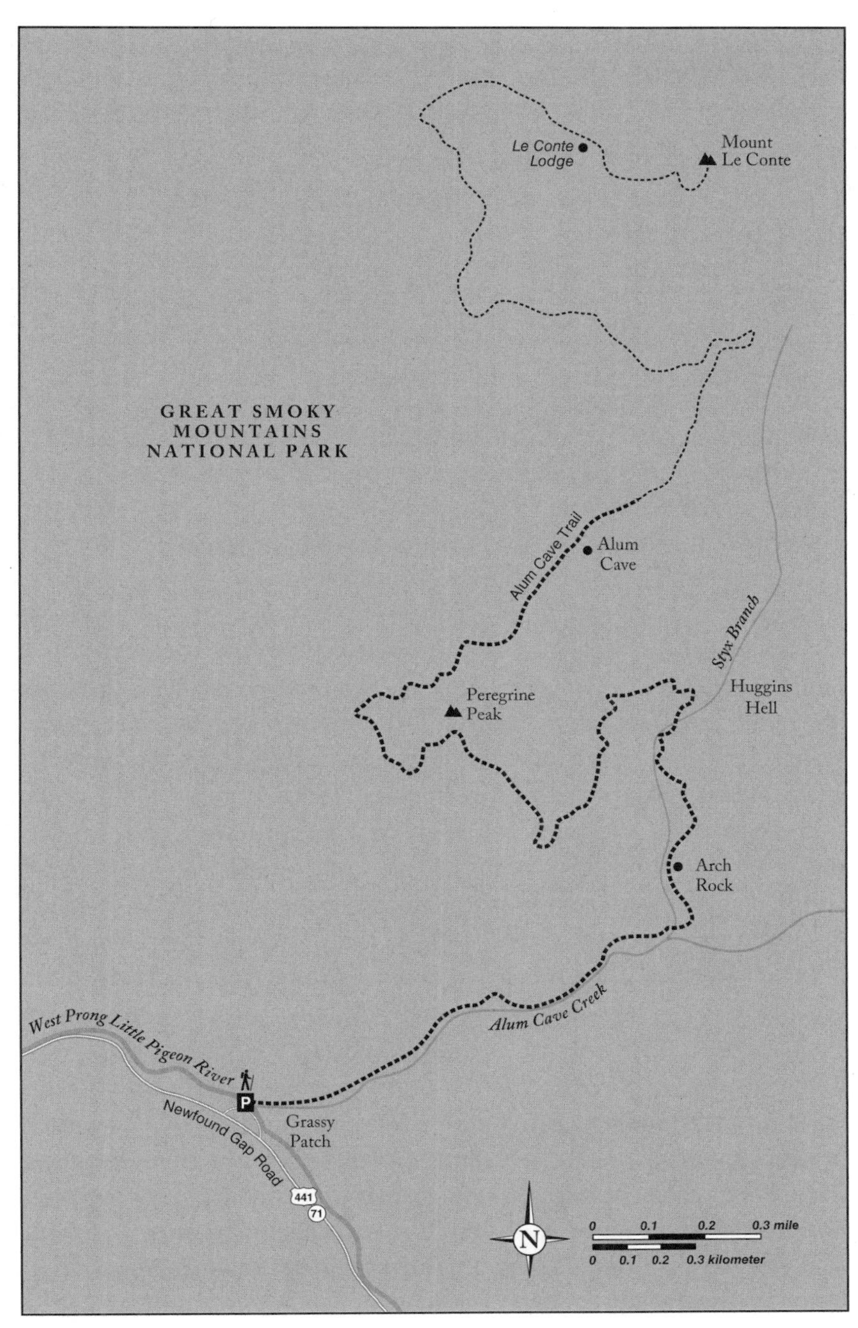
Le Conte
Lodge
Mount
Le Conte
GREAT SMOKY
MOUNTAINS
NATIONAL PARK
Alum Cave Trail
Alum
Cave
Styx Branch
Peregrine
Peak
Huggins
Hell
Arch
Rock
Alum Cave Creek
West Prong Little Pigeon River
P
Grassy
Patch
Newfound Gap Road
441
71
N
0
0.1
0.2
0.3 mile
0
0.1
0.2
0.3 kilometer

17. Alum Cave

17. ALUM CAVE

Highlights: Walk along a rushing mountain stream, then climb past rugged rock features to outstanding views of steep slopes and ridges, some supporting extensive shrub balds. Eventually you'll reach the unusual overhanging rock formation of Alum Cave.

Natural Communities: Northern hardwood forest, cove forest (acidic subtype), shrub bald

Elevation: 3,900 ft (trailhead); 4,900 ft (Alum Cave)

Distance: 4.5 mi round-trip

Difficulty: Moderate to strenuous, with an elevation gain over 1,000 ft in 2.25 mi

Directions: From Newfound Gap, follow US 441 west into Tennessee and park at one of two adjacent trailhead parking areas about 4.3 mi on the right. If approaching from the Gatlinburg, TN, area, take US 441 about 8.3 mi into the park from the Sugarlands Visitor Center; trailhead parking areas will be on your left (no facilities).

GPS: Lat. 35° 37′ 43.8″ N; Long. 83° 27′ 3.2″ W (trailhead)

Information: Great Smoky Mountains National Park, 107 Park Headquarters Rd., Gatlinburg, TN 37738, (865) 436-1200

hell," and you'll agree with them if you attempt to travel any distance off-trail through this dense tangle. Later in the hike, you'll veer away from Alum Cave Creek along its tributary, Styx Branch. The latter is aptly named, given its origins in Huggins Hell!

Further along the trail you'll encounter reminders that related high-elevation forest communities are close at hand. You'll see an occasional yellow buckeye (*Aesculus flava*) from the northern hardwoods forest and red spruce (*Picea rubens*) from the spruce-fir forest. The presence of large red spruce trees as low as 4,000 ft may surprise you, especially if you have hiked to Gregory Bald, which lacks red spruce even at its 5,000-ft summit. One explanation for these lower-elevation occurrences of red spruce in the heart of the Great Smoky Mountains is that the adjacent high peaks served as a refuge for red spruce during an earlier period of warmer climate; spruce then migrated downslope when the climate cooled again. Meanwhile, outlying high peaks such as Gregory Bald may have lost their red spruce during that warmer period and never regained them because of the great distance to nearby populations. You may enjoy reading "The Case of the Missing Spruce" sidebar in hike 3 (Brasstown Bald) for more information on this topic. Presence of red spruce at relatively low elevations along the Alum Cave Trail may also have a climatic explanation. Sites in the interior of a large mountain range (such as the Alum Cave Trail) have more "montane" conditions (cooler, moister) than their elevational counterparts on the periphery of the range (like Gregory Bald). German ecologists call this phenomenon the *massenerhebung* effect.

At about 1 mi along the trail and 4,200 ft elevation, you'll encounter a rustic log bridge across Alum Cave Creek. Pause here to look closely at the riparian vegetation, particularly the structure of the abundant yellow birch. You may notice that large patches of yellow birch are comprised of relatively small stems of the same diameter. These notably young and even-aged stands likely regenerated after catastrophic floods that occasionally sweep through these creeks, typically after summer storms in the high peaks.

One such deluge on September 1, 1950, sent floodwaters and debris raging through this typically peaceful valley, scouring the valley and tearing out the forest. Upslope, massive rockslides tumbled, and you may have noticed their persistent scars on the ridge slopes if you approached the Alum Cave parking lot from Newfound Gap. Additional flood events occurred in the 1970s and in 1993, so you may also note yellow birch patches of distinctly different ages. Our sidebar for this hike discusses yellow birch, which often benefits from such devastating events.

After a third creek crossing on a log bridge, you will reach Arch Rock, a massive boulder almost blocking the trail. This layered metamorphic rock has fractured over time to offer the hiker a steep and narrow passage through its interior. Aided by carefully placed stone steps and steel cables, you can climb up and through Arch Rock to reach the trail above. After a final crossing of Styx Branch on a log bridge, you will find yourself climbing steeply, hugging the rock wall as you make the final ascent to the shrub bald above. Pause to catch your breath and observe the large red spruce, some up to 2 ft in diameter, along this stretch of the trail.

Keep hiking, and you'll notice that the stature of the forest is becoming shorter and shorter. As the trees give way to rhododendrons, you may feel like Alice in Wonderland after partaking of an "EAT ME" granola bar—it's as if you've now grown to the stature of the canopy itself. In reality, you've just crested a sharp ridge, and you're immersed in a shrub bald only slightly taller than you are. Continue

Yellow Birch—Engineered for Disaster

If you are familiar with the life history of yellow birch (*Betula alleghaniensis*), you will understand why it is so common in sometimes-tumultuous stream corridors. Birch trees produce abundant wind-borne seeds that germinate well on exposed mineral soil. The seedlings grow quickly in full sun, easily outpacing slower growers like eastern hemlock (*Tsuga canadensis*) and red spruce (*Picea rubens*). The mineral sediments exposed and deposited following landslides and floods provide abundant resources and a perfect environment for an opportunist like yellow birch, so it is common at middle to high elevations in areas that experience occasional catastrophic disturbance.

In addition, one of yellow birch's strengths is its ability to germinate and establish itself on moss-covered boulders and decaying logs. Taking advantage of the abundant light in a gap above, seedlings scavenge whatever moisture and nutrients they can from moss beds or decaying wood. As the seedlings grow into saplings, they send out long roots that snake across and around their log or boulder hosts, eventually (if successful) reaching mineral soil with its reliable supply of moisture and more abundant nutrients. Thus anchored, a sapling can grow to tree size, still clasping its host rock or log in an octopuslike grip. You can find large yellow birches in the vicinity of this stop still holding tight to their boulders. More interesting, however, are the yellow birches that began life on large decaying logs. These trees still faithfully record, in the remarkable shapes of their gnarled stilt roots, the dimensions of their original host logs, long since rotted away.

Yellow birch appears to stand on tiptoe over the remains of the log where it first germinated as a tiny seedling. This ability to send roots around rocks and logs and into the soil enables yellow birch to successfully colonize disturbed areas.

to a sharp bend in the trail, where you'll find an inviting rock outcrop that has been used as a resting place by generations of hikers. Select a safe perch and take in the view, if you're fortunate enough to have arrived on a clear day. To the west, you can see down the valley of the West Prong of the Little Pigeon River. The nearest ridge across the abyss is razor-sharp, and if you scan its crest, you may spot a place where weathering has carved a human-sized hole through the rock. To the east, you can look back up the valley toward Newfound Gap, and you'll see that some of the nearby ridges support shrub balds similar to the one in which you're located.

The shrub bald you've reached is hardly bald, although it may look that way from far below. The steep slopes are cloaked in a dense thicket of shrubs, most of them in the Ericaceae (blueberry) family. You should be able to find three species of evergreen rhododendrons (rosebay, Catawba, and gorge [*Rhododendron minus*]). Mountain laurel is also abundant, and along the rocky edges of shrub thickets, you'll find its close relative, sand myrtle (*Kalmia buxifolia*). Blueberries are common here, and you'll also find a variety of non-ericaceous shrubs like the withe-rod (*Viburnum cassinoides*), along with stunted trees such as red maple (*Acer rubrum*) and red spruce. Larger trees are uncommon but do occasionally poke up through the shrubs.

The shrub bald is just below 5,000 ft, far below the climatic treeline at this latitude. Why, then, are you in a shrub-dominated community? The answer is beneath your feet! The rock you're standing or sitting on is the Anakeesta Formation, an easily fractured, acidic metamorphic rock with its beds tilted steeply from horizontal. As this rock weathers, it forms shallow, unstable soils and steep slopes, a perfect situation for rockslides and a terrible place for trees to establish a firm foothold. The trees that do grow here are readily thrown over in high winds, and the soil is constantly slipping beneath their roots. In contrast, ericaceous shrubs thrive here; their extensive root systems grip the shallow soils, and they prefer acidic, well-drained soils. Plenty of rainfall and sunlight complete the list of requirements for an ideal shrub bald. In a study of shrub balds throughout the southern Appalachians, Peter White of the North Carolina Botanical Garden discovered that their existence was predictable, given conditions similar to the very ones that you find in this corner of the Smokies.

Just a quarter mile beyond the shrub bald overlook (and 2.25 mi

From your vantage point on the shrub bald, look east across the valley to see how another shrub bald looks from afar. Notice the flattened appearance of the heaths on the ridgeline and steep flanks below, and what are most likely evergreen red spruces creeping into the ravines and lower slopes.

from the trailhead) is the namesake of this trail, Alum Cave. Despite its name, the massive rock feature looming above and to your right is not a cave but, rather, a gigantic rock overhang sheltering an area the length of a football field. The rock here weathers to a fine powder that has accumulated beneath the overhang. The sulfurous minerals were rumored at one time to contain saltpeter, an ingredient in the manufacture of gunpowder. In reality, the only mineral commercially mined here was Epsom salt, extracted during the 19th century. Far above, you can often see water dripping from the lip of the rock overhang, and you might imagine the massive icicles that form here in winter.

Two uncommon birds can be seen plying the airspace surrounding these high cliffs. Ravens (*Corvus corax*), somewhat larger than crows, can be identified by their trademark "gronk" calls; they are a delight to watch as they soar and tumble on the thermals. Also occasionally seen here are peregrine falcons (*Falco pererinus*), small raptors that are the speed champs of the animal world, capable of reaching 200 mph as they dive-bomb their avian prey from high above. Once nearly brought to extinction by the harmful effects of DDT, peregrines are now breeding successfully here in the high Smokies.

When you've finished resting in the shelter of Alum Cave, you have two hiking options. Our recommended route for a half-day trip is to return the way you came to the trailhead. Hikers prepared for a longer adventure may continue to Mount LeConte, another 2.75 mi ahead and 1,400 ft of climbing. If you opt for Mount LeConte, please consult one of several available hiking guides for trail information. The round-trip hike from the Alum Cave trailhead to Mount LeConte and back makes for a long and strenuous day hike, covering 10 mi and 2,500 ft elevation gain and loss.

Regardless of how you spend the rest of your day on the Alum Cave Trail, if you pause in the shrub bald on your return hike, think about how the rocks beneath began their existence as sediments deposited in a shallow sea 600 million years ago. An unimaginably long history of sedimentation, mineral deposition, and metamorphosis under high temperature and crushing pressure created these rocks, which were then heaved and tilted as they were lifted from their oceanic birthplace to great heights during the formation of the Appalachians. The unique properties of these rocks provide habitat for the unusual and lovely shrub bald we enjoy today.

A half-hour spent resting by the rushing cascades of Alum Cave Creek under the shelter of a yellow birch canopy is also time well spent. Here you might think of those rare but important times when a deluge on the peaks above sends floodwaters raging through this valley, scouring it with uprooted trees and tumbling boulders, leaving in place sediments that will support the next generation of yellow birch and other opportunists. We live in a world of change, sometimes imperceptibly slow and sometimes speeding like a runaway locomotive. The forests and shrub thickets of these mountains are both witnesses to and beneficiaries of this change, and we are the richer for the biological diversity they offer as we hike through these ancient and beautiful mountains.

18. RAMSEY CASCADES

Why Are We Here?

Ramsey Cascades is a popular hike in the quieter Greenbrier section of Great Smoky Mountains National Park. The trail traverses an elevation gradient, transitioning through different forest types as it climbs toward the cascades. There is also a disturbance gradient along the trail, which traverses secondary to old-growth forest. We'll focus on this gradient of disturbance history, established when loggers harvested the first, level section of the forest along the Middle Prong of the Little Pigeon River but never reached the upper portions. The cascades themselves make for a rewarding destination, offering a perfect lunch and rest stop after a long, ascending hike.

The Hike

From the parking area beside the Little Pigeon River, follow an old road to start your hike. This road is also the trail for the first mile and a half, winding through a second-growth forest. You can tell that this forest is second growth by the relatively small and uniform sizes of the trees. When the park was established in 1934, only 20% of its forests remained intact as old-growth. Stands like the one you find near the start of this trail had already been clear-cut, so the trees are only slightly older than the park itself. You can look for the change in harvest history and practices as you climb. The most easily accessed areas, where you start your hike, were cleared almost entirely, the most valuable trees were selectively logged farther up, and the remaining miles to the cascades were left in a pristine condition.

The canopy dominants are tulip-tree (*Liriodendron tulipifera*) and eastern hemlock (*Tsuga canadensis*). Both are common species in the southern Appalachians and found in many natural communities, so their presence doesn't help you classify the forest. Closest to the stream, though, you will also see the stark white, peeling park of American sycamore (*Platanus occidentalis*), an uncommon tree in the mountains and found only along streams at lower elevations. As you climb, you'll notice that this tree disappears from the forest quickly.

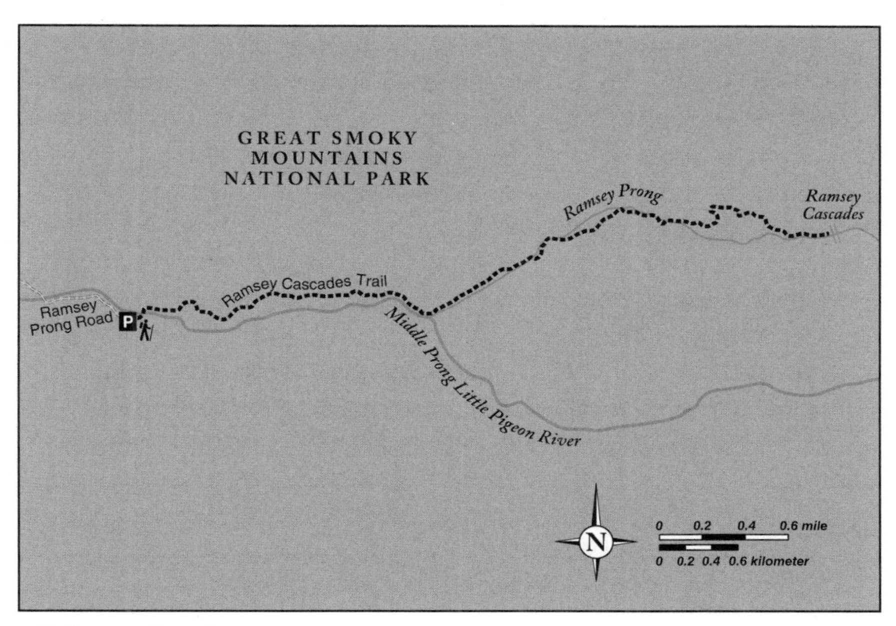

18. Ramsey Cascades

The stream banks are covered with rosebay rhododendron (*Rhododendron maximum*). The presence of rosebay rhododendron and other shrubs in the blueberry family, like mountain laurel (*Kalmia latifolia*) in the drier areas and galax (*Galax urceolata*), with its round, shiny leaves, tells you that the cove forest you're walking through is acidic. Furthermore, the trees that grow along the creek must be adapted to moisture and humidity. Hemlock and rhododendron thrive in these moist environments, and they are superior competitors here, so they dominate.

Farther from the stream, other abiotic factors come into play, and the natural community becomes a cove forest. You find a cove forest in this mid-elevation site because of its relative protection from harsh weather by the bowl-shaped topography. Deep soils also accumulate here, adding to the moisture storage capacity of the system. As you hike, however, you'll also notice two variants of the cove forest—acidic and rich.

Why is there so much compositional variation in such a small space? It is unlikely that the soil parent material (underlying rock) exhibits that much variation. However, subtle moisture differences

18. RAMSEY CASCADES

Highlights: A beautiful climb through a continuously changing forest to a roaring cascade

Natural Communities: Cove forest (both acidic and rich subtypes), northern hardwood forest

Elevation: 2,147 ft (trailhead); 4,400 ft (Ramsey Cascades)

Distance: 8 mi round-trip to Ramsey Cascades

Difficulty: Moderate walking at first, then a steady climb to the cascades. The descent on the way back is easy.

Directions: From Gatlinburg, TN, take US 321 east for nearly 6 mi. Turn right on Greenbrier Rd. and drive 3.2 mi, past the ranger station, where the road turns to gravel. Turn left at the sign for Ramsey Cascades across a bridge. Parking is at the end of this road, about 1.5 mi farther (no facilities).

GPS: Lat. 35° 42′ 9.6″ N; Long. 83° 21′ 25.9″ W (trailhead)

Lat. 35° 42′ 50.5″ N; Long. 83° 18′ 1.0″ W (Ramsey Cascades)

Information: Great Smoky Mountains National Park, 107 Park Headquarters Rd., Gatlinburg, TN 37738, (865) 436-1200

might allow some species, like hemlock and rhododendron, to dominate in certain areas, beginning a feedback loop where their dense growth and acidic leaf litter mean that few other species can thrive.

In the areas where tulip-tree dominates, there is a better-developed herb layer, as well as greater diversity among the tree species. Other typical rich cove tree species are present, including sugar maple (*Acer saccharum*), northern red oak (*Quercus rubra*), basswood (*Tilia americana*), and yellow buckeye (*Aesculus flava*). You might also notice an occasional tree with smooth leaf edges and deep blue/black flaky bark with orange streaks; this is a cove specialist, Carolina silverbell (*Halesia tetraptera*), and if you are here in late spring, you might see its white, bell-shaped flowers scattered across the forest floor, prompting you to peer above your head to see the source of these beauties.

For the first mile and a half, you'll follow the old roadbed, with the music of the stream as your constant companion. You'll come to a section running close to the stream for the next mile, and you would expect this forest to be dominated by hemlock and rhododendron. However, the hemlock trees here are dead and falling down, leaving large gaps in the forest canopy. The sidebar offers the details of this tragic tale.

The fallen hemlocks create large gaps in the canopy, allowing sunlight to reach the forest floor of what was previously a deep, dark forest. Pause by several of these openings and look carefully to see if you can determine which trees will eventually fill these gaps. In some places, it seems that the dense rhododendron already in place near the stream will best exploit the increased light. Their thick foliage reduces the amount of light reaching the forest floor and prevents germination by other species.

In other places, you can see that one tree has clearly captured a canopy gap; in still other places, there is a race between two or more trees to capture the available light and other resources. In many places at lower elevations, you can see red maple (*Acer rubrum*) taking over these spaces. Red maple is a generalist species that has an advantage here. It can tolerate some shade and very wet as well as rather dry conditions. Red maple also grows quickly, enabling it to fill the newly created space and shade competitors. Some ecologists think that red maple is slowly taking over eastern forests because fire, its only limitation, has been suppressed since European settlement. Meanwhile, many other tree species have become more susceptible to disease and/or pests. The absence of big trees along the lower section of the trail indicates that this area was logged before it became part of the park; you might also notice areas of even-aged tulip-trees, which seeded in naturally after harvest.

The trail narrows at around 1.5 mi as the old road becomes a trail. At this point, logging must have switched from large-scale clear-cutting to selective harvest of the best and most valuable trees, so you may begin to notice some larger individuals of less commercially valuable species, such as tulip-tree. You'll also find that the trail becomes steeper and narrower as you begin to climb, and you'll cross the stream several times on your way up. At mile 2.5, the trail passes two enormous tulip-trees and enters an older-growth stand untouched by the big saws of loggers long ago. Not only will you see large trees, but you'll see trees of different sizes along the way. You may also notice that areas without rosebay rhododendron are richer in canopy species, accompanied by a better-developed herbaceous layer with more wildflowers.

If your aching leg muscles haven't convinced you already, there are other indicators that you are gaining elevation as you climb toward

Hemlock Woolly Adelgid

Eastern hemlocks (*Tsuga canadensis*) (as well as Carolina hemlocks [*T. caroliniana*]) are dying en masse from effects of a deadly, nonnative insect, the hemlock woolly adelgid (*Adelges tsugae*). This invasive, aphidlike insect from eastern Asia was discovered near Richmond, VA, in the 1950s, and the insect dispersed readily, spread by birds and wind and piggy-backing on horticultural specimens. By the mid-1990s, it had spread into the southern Appalachian Mountains, reaching Great Smoky Mountains National Park by the early 2000s.

Treatment methods, which include insecticide applications and biological control, have been locally effective but are impractical to implement on a large scale. Look carefully at living hemlocks you encounter while hiking; the adelgid can be easily seen at the base of the hemlocks' short needles, usually evident from the fuzzy white coating it produces to protect itself and its eggs. The insect feeds on the tree's sap, blocking nutrient flow and causing the tree to die within 3 to 5 years. Small groves of healthy trees you might see at park visitor centers and other prominent places are receiving ongoing insecticidal treatment.

In the meantime, large swaths of hemlocks have perished. Hemlock has been described as a foundation species for some natural communities, including acidic coves. A foundation species is one that creates structure or provides ecological services for many of the other species in its natural community. Here hemlocks provide important winter shelter for many bird species, and they create deep shade that keeps mountain streams cool

the cascades. Higher-elevation species begin to appear, such as witch hobble (*Viburnum lantanoides*), a tall shrub with large, paired, heart-shaped leaves, more often found in northern hardwood forests. In fact, as you look around, you'll notice that the canopy dominants have changed from tulip-tree to species such as yellow birch (*Betula alleghaniensis*), sugar maple, and yellow buckeye. The lower-elevation rosebay rhododendron is joined by its high-elevation relative, Catawba rhododendron (*Rhododendron catawbiense*), with its magenta flowers. Eastern hemlock continues its dominance along the creek, but you might see a shift in the species coming in to replace fallen trees.

Hemlock woolly adelgid can be identified by the fuzzy white coating on insects and eggs attached to the bases of the needles of both eastern and Carolina hemlock. Once an infestation occurs, the trees die within 3 to 5 years.

for species like the native brown trout (*Salmo tratta*). Long-term concerns related to the demise of the hemlocks include negative consequences for these wildlife populations, in addition to the effects of large amounts of dead wood and debris that will accumulate as vast numbers of trees are felled by this deadly insect. It's also unclear which species may eventually fill the hemlock's niche.

You'll hear Ramsey Cascades before you see them, and you may be surprised, if you journeyed here on your own, to see many other hikers enjoying the falls. Find a good resting spot to enjoy a well-deserved break, but keep your eye on your snacks! Territorial red squirrels (*Tamiasciurus hudsonicus*) have strategically staked out the area around the cascades, and they are quite bold. Listen for their chirrups and watch their antics as you rest. Though widespread across northern North America, these smaller squirrels with reddish tails and cheeky attitudes reach the southeasternmost part of their native range here in the southern Appalachians. Residents of boreal forests,

After crossing into the old-growth part of the rich cove forest at Ramsey Cascades, look up to admire the enormous spreading canopies of the biggest tulip-trees.

they likely supplement their conifer seed diets with granola bars here at Ramsey Cascades. No matter how captivating you might find them, however, don't be tempted to feed them—they are wild, and human food should not be part of their diet.

As you rest, look around at the forest surrounding the cascades. Dead and dying eastern hemlock are still here, along with healthy yellow birch. You are above 4,000 ft near the base of the cascades, and you will see that evergreen red spruce (*Picea rubens*) has joined the list of canopy species. You are indeed sitting in a transition zone between northern hardwood forest and spruce-fir forest.

The occurrence of a transition, or an ecotone, between different natural communities is quite evident on this hike. By following Middle Prong up to the cascades, you have maintained a fairly consistent moisture level along the 4-mi trail, though you've also noticed subtle differences in vegetation as you moved closer or farther from the stream. However, you have now climbed well over 2,000 ft on your way to Ramsey Cascades, and you have seen the forest transition from acidic/rich cove to northern hardwood forest to spruce forest. Along this trail the transition is particularly subtle; rather than an abrupt shift in vegetation, you may have noticed certain trees present at the beginning but absent farther up. You may also have noticed a shift in the amount of vegetation in the understory. The herb layer you gained in the middle part of the hike diminished again as you reached the upper section. The gradual transitions in composition of natural communities along environmental gradients prompted some ecologists in the early 20th century to question the dominant paradigm of communities as discrete and highly organized entities. To learn more about this debate, please read the "Community Concepts in Conflict" sidebar in hike 15 (Gregory Bald).

If you didn't study the shift in forest species on your way up, your descent now affords a great opportunity to observe transitions in forest cover. Choose an easily recognized species, such as red spruce, witch hobble, yellow birch, or tulip-tree, and, as you descend, watch for your chosen woody plant to shift its abundance along your way (this is even more fun if everyone in your party chooses a different species, allowing you to compare notes). Take note of the first time you see your tree—how abundant is it?—then, as you descend, pause and assess its abundance again at various points. Another way to observe this phenomenon is to examine a gap caused by a fallen eastern hemlock and then determine which species seems poised to replace it. While red maple was a common contender at lower elevations, it is likely that a different species is a better competitor in the upper elevations along this trail. Recognizing the subtle distinctions among forest communities and how they change along an elevation gradient like this one allows us to understand landscapes as a whole.

19. CRAGGY PINNACLE

Why Are We Here?

Craggy Gardens, including Craggy Pinnacle, is one of the most visited areas along the Blue Ridge Parkway in North Carolina. Many people

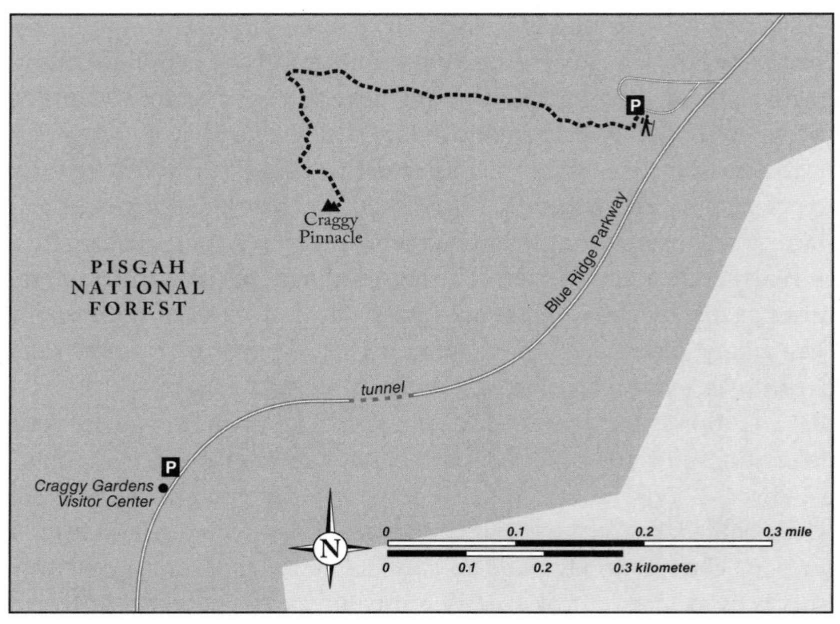

19. Craggy Pinnacle

come to the Craggy Mountains to see Catawba rhododendron (*Rhododendron catawbiense*) at its peak floral display in late June and early July. However, Craggy Gardens is a delightful place to visit at any time from late spring to early fall, and the Craggy Pinnacle Trail is a convenient way to experience the "gardens" up close, with rewarding views at the summit of some of the highest peaks in eastern North America. Craggy Gardens is also an excellent place to reflect on the dynamic nature of vegetation, as influenced by the physical environment, past history, and animals (including humans).

The Hike

The Craggy Pinnacle Trail offers an out-and-back hike from the Craggy Dome overlook/parking area to the summit of Craggy Pinnacle. The hike begins at the upper parking area, to the right of two interpretive signs. The first part of the trail passes through a stunted northern hardwood forest where the dominant species is yellow birch (*Betula alleghaniensis*). Other tree species that you may see here include American beech (*Fagus grandifolia*), yellow buckeye (*Aesculus flava*), mountain

19. CRAGGY PINNACLE

Highlights: A moderate hike through northern hardwood forest and heath bald, ending at a rock outcrop with 360° views of the Craggy and Black Mountains; opportunities to see rare plants; spectacular floral display of Catawba rhododendron in June and July

Natural Communities: Northern hardwood forest, shrub bald, rocky summit

Elevation: 5,640 ft (trailhead); 5,892 ft (summit)

Distance: 0.7 mi to the summit from the trailhead; 1.4 mi round-trip

Difficulty: Moderate; some steeper, wet, and rocky spots, but mostly easy going

Directions: From Asheville, take the Blue Ridge Parkway 24 mi north, past the Folk Art Center, to the Craggy Gardens Visitor Center (MP 364.6) (restrooms, information, gift shop). Continue 0.5 mi north on the Blue Ridge Parkway, passing through Craggy Pinnacle tunnel. Parking for the Craggy Pinnacle trailhead will be shortly after the tunnel at the Craggy Dome Overlook on your left (MP 364.1).

From the Mount Mitchell area, head south on the Blue Ridge Parkway about 10 mi, and turn right into the parking area for the Craggy Dome Overlook (MP 364.1). This section of the Blue Ridge Parkway is closed during November and December and at other times when unfavorable road conditions warrant.

GPS: Lat. 35° 42′ 12″ N; Long. 82° 22′ 40″ W (summit of Craggy Pinnacle)

Information: Blue Ridge Parkway Headquarters, 199 Hemphill Knob Rd., Asheville, NC 28803-8686, (828) 271-4779 (main line), (828) 298-0398 (park information line for facilities, activities, and road closures)

Blue Ridge Parkway Association, P.O. Box 2136, Asheville, NC 28802-2136, (828) 670-1924

ash (*Sorbus americana*), and hawthorn (*Crataegus flava*). The prevalence of yellow birch here is unsurprising, given the rockiness of the terrain and yellow birch's propensity to establish a foothold and grow over rocks and boulders at higher elevations (see the "Yellow Birch—Engineered for Disaster" sidebar in hike 17 [Alum Cave]). Some of the yellow birches along the trail are large and apparently old, based on their husky stature, yet they rarely exceed 20 ft in height, a testament to the severity of winter weather at this elevation. There are a couple of especially gnarly and photogenic yellow birches along the first stretch of the trail.

Given the elevation of this site (well over a mile high), one might expect to find a well-developed spruce-fir forest. The entrance to Mount

Mitchell State Park, in the heart of the Black Mountains, is just 10 mi north on the Blue Ridge Parkway, and you can readily find spruce-fir forest there. If you have completed some of our other hikes (Gregory Bald in the Great Smokies [hike 15] and the Appalachian Trail on Roan Mountain [hike 25]), you will have encountered this paradox before: absence of spruce-fir forest at an elevation where you would expect it. Here in Craggy Gardens, we can rule out logging as a possible explanation. However, we cannot rule out a devastating presettlement fire, or it is possible that this area has been open alpine grassland since the last glacial maximum and has only recently been reforested or overtaken by heaths. This is a problem that we'll discuss later, but first let's hike!

The trail eventually winds around a large boulder, and above this point, the canopy is comprised mostly of ericaceous shrubs, with Catawba rhododendron, mountain laurel (*Kalmia latifolia*) and blueberries (*Vaccinium* spp.) most prominent. Interestingly, the understory has a rich herbaceous layer of perennial herbaceous species that are common at this elevation; this is a bit surprising, since these herbs are often suppressed beneath evergreen shrubs. One might wonder if these herbs persist from a recent time in the past, when this area was much more open. As you climb toward the summit, the shrubby canopy becomes increasingly gnarled and stunted. The upper overlook (take the right fork of the trail near the summit and proceed to the walled area) is a good place to begin your visit to Craggy Pinnacle. You can then continue your exploration of the summit at the lower overlook (left fork of the trail just before reaching the summit).

The summit view is spectacular (weather permitting, of course). Looking back the way you came (to the east), the closest peak you'll see, rising approximately 1,000 ft above the surrounding terrain, is Craggy Dome (summit elevation 6,090 ft). Beyond, to the northeast, are the Black Mountains, with Mount Mitchell towering above. To the southeast, you can look down toward Asheville's water supply, the North Fork (Burnett) Reservoir. Immediately to the south, and far below you now, is the Craggy Gardens Visitor Center, and beyond you can see the Blue Ridge Parkway headed south. If the air is clear, you'll also see the outskirts of Asheville in the distance. If the air is *very* clear, you may see the Great Smoky Mountains, far to the southwest.

After taking in the distant views, consider the rhododendron gar-

Hikers are rewarded with panoramic views from the summit of Craggy Pinnacle on a clear day. Carefully designed barriers protect both visitors from falls and rare plants from trampling. Peek over the edge to see if you can glimpse some of the rare glacial relicts that call Craggy Pinnacle home.

den in the foreground. Although it's difficult to imagine, the heath bald surrounding you was a grassy meadow in the early part of the twentieth century, as well-documented in a photograph taken by W. A. Barnhill in 1915. Interestingly, this photograph shows a herdsman salting a flock of sheep on Craggy Flats below an open Craggy Pinnacle; scattered trees (both live and dead) suggest that the grassy area of that time had been expanding at the expense of forest.

Many would argue that the remaining grassy meadows or grass balds of the southern Appalachians exist today only under a frequent grazing or mowing regime, and they may have been maintained that way since the end of the Pleistocene by native grazers and, eventually, by domestic cattle and sheep brought in by European settlers. Without the grazers and browsers, these grassy areas have quickly filled in with native shrubs and, in some places, seem headed toward a takeover by northern hardwood forest. Would you advocate managing this area for open space, as is the case on the Roan Massif (refer to hike 24 [Roan High Bluff and Roan Gardens])? Or is the change that has occurred here in the past century simply nature's way, which

A grassy Craggy Pinnacle is seen in this historic photograph taken by W. A. Barnhill. The question is this: What role did domestic grazing animals play in creating grass balds and/or maintaining them in an open condition? (Herdsman salting his sheep. Great Craggy Mtns, Craggy Pinnacle [left] and Craggy Knob [right] background. Barnhill photo numbered B-673, date 1914–17. Copy print by Barnhill.) Photograph credit: William A. Barnhill Collection, Pack Memorial Public Library, Asheville, North Carolina.

should be left alone? Your answers to these questions may be influenced by our discussion of rare plants that follows!

As you contemplate the management of Craggy Pinnacle, walk around the summit within the confines of the stone walls and examine the rocky summit. This is one of the best places in the southern Appalachians to see rock outcrop plants that were likely widespread during the most recent glacial maximum (18,000 years before present), when all the surrounding terrain would have been alpine tundra. These plants still persist as "glacial relicts," despite the present moderate climate that has allowed forest and heath to cover most of the former alpine habitat. This might be a good time to read the "Ice Age Relicts" sidebar in hike 11 (Satulah Mountain).

Look for three-toothed cinquefoil (*Sibbaldia tridentata*), which appears very much like a miniature red-leaved strawberry plant when its 5-petaled white flowers are in bloom. Another arctic-alpine species

Loving a Place to Death

Please stay on the trail or within the confines of the walled overlook areas on the summit at all times during your Craggy Pinnacle hike. Craggy Pinnacle has one of the densest concentrations of rare plants in the southern Appalachians, many of which can be (and have been) inadvertently trampled by off-trail hikers, photographers, and blueberry-gatherers. Not only is it ethically unacceptable to leave the trail for any reason, but it is also illegal!

Even well-intentioned botanists have contributed to the decimation of high-elevation rare flora in the southern Appalachians. Consider Alan Weakley's note about cliff avens (*Geum radiatum*) in *Flora of the Southern and Mid-Atlantic States* (2001): "It is illegal to collect *G. radiatum* without federal and state permits, and there is no justification (scientific or otherwise) for additional collections from known sites. This is one of the few plant species that has been seriously depleted by collection by scientists (several hundred herbarium sheets from Roan Mountain alone!), though recreational over-use of its habitats, and possibly also pollution and break-up of adjoining spruce-fir forests, are the most critical threats to its continued existence."

Needless to say, the National Park Service, charged with protecting both native species and their habitats along the Blue Ridge Parkway, is concerned about the protection of sensitive rare-plant sites like Craggy Pinnacle. The walled overlook on Craggy Pinnacle was actually designed by a graduate student at the University of Georgia to facilitate protection of rare plants, but visitors still persist in hiking beyond this barrier. As one consequence of this degradation of a precious environmental resource, the 2013 Blue Ridge Parkway Management Plan calls for the eventual closure of the Craggy Pinnacle Trail to the public to safeguard the rare plants and to allow the highly eroded trail to recover.

You may thus find the Craggy Pinnacle Hike unavailable if you visit the Craggy Mountains. Fortunately, alternative similar hikes in the area do exist (one could hike the trail to the Craggy Gardens Picnic Area from the Craggy Gardens Visitor Center, for example). If you are able to hike to Craggy Pinnacle, please consider the difficulty of managing a place such as this, where imperiled species coexist with ongoing human visitation.

in the rose family is a rare and federally endangered plant, the yellow-flowered cliff avens (*Geum radiatum*). Unlike three-toothed cinquefoil, this species is an endemic, found only on a few high peaks in North Carolina. Finally, look for the Appalachian firmoss (*Huperzia appressa*, a member of the clubmoss family, a primitive plant group distantly related to ferns), a southern Appalachian disjunct plant with a widespread distribution in the boreal regions of the globe.

Many of the other rare, glacial-relict plants here are grasses or grasslike plants and considerably more difficult to identify without close inspection. The most distinctive of these is the deerhair bulrush (a member of the sedge family) (*Trichophorum cespitosum* ssp. *cespitosum*; the "ssp." means "subspecies"), which can be found growing in crevices and on small ledges among the rocks, with thick and drooping clumps of persistent dried leaves. Deerhair bulrush is also a disjunct species, with scattered populations in the southern Appalachians and a broader distribution in alpine and arctic habitats across northern North America and Eurasia. The same can be said of Highland rush (*Juncus trifidus*), a small, grasslike plant that can also be found growing in rock crevices here. A final noteworthy plant (this time a true grass) is Cain's reed-grass (*Calamagrostis cainii*), a rare and narrow endemic whose entire global distribution is Craggy Pinnacle and only two other mountain summits in the southern Appalachians.

One of the most serious threats to the rare plants of high-elevation rock outcrops and rocky summits in the southern Appalachians is traffic—not of automobiles, but of *Homo sapiens*. We discuss this threat at greater length in the sidebar.

Climate change is also a foreboding challenge for rare species of high elevations in the southern Appalachians, because a warming climate will push many of these species off their remaining mountaintop habitats (the so-called sky islands) forever. For the disjunct species, such as three-toothed cinquefoil, deerhair bulrush, highland rush, and Appalachian firmoss, other populations far to the north would presumably remain. For the endemic species (cliff avens and Cain's reed-grass, for example), loss of their southern Appalachian mountaintop refuges would spell almost certain extinction. Like the woolly mammoth (*Mammuthus primigenius*), these species would join the legions of other species that could not sustain the dual postglacial onslaughts of climate change and burgeoning human populations.

You may read more about this concern in the sidebar "Climate Change on Sky Islands" in hike 23 (Roan High Bluff and Roan Gardens).

20. MOUNT CRAIG

Why Are We Here?

"The woods are lovely, dark, and deep
But I have promises to keep."

You might recall these words of the New England poet Robert Frost as you descend the steep stone stairway to begin your hike through the dense forest along the ridgeline of the Black Mountains. Indeed, this is entirely appropriate, because the trees you see on your walk today have close ties to species in northern New England and Canada. This hike gives you not only a feeling of the climate and landscape of points far to our north but also a glimpse into how the spruce-fir forest may have appeared in the earlier part of the 20th century, before the arrival of the exotic invasive insect, the balsam woolly adelgid (*Adelges piceae*).

Mount Mitchell has the distinction of being the highest peak east of the Rocky Mountains and the site of the first state park in North Carolina. It was named for University of North Carolina geography professor Elisha Mitchell, who fell to his death on a frozen waterfall while completing his measurements to demonstrate that the elevation of this lofty summit exceeded that of Clingman's Dome in the Smokies. When you return from your hike to Mount Craig, you will certainly want to complete the short hike up to the summit of Mount Mitchell and its observation tower. Then, visit the excellent museum and enjoy a cup of hot chocolate from the concession stand.

The Hike

Descend on the paved path and follow the gravel trail through the picnic area to the start of the Deep Gap Trail, marked by white trian-

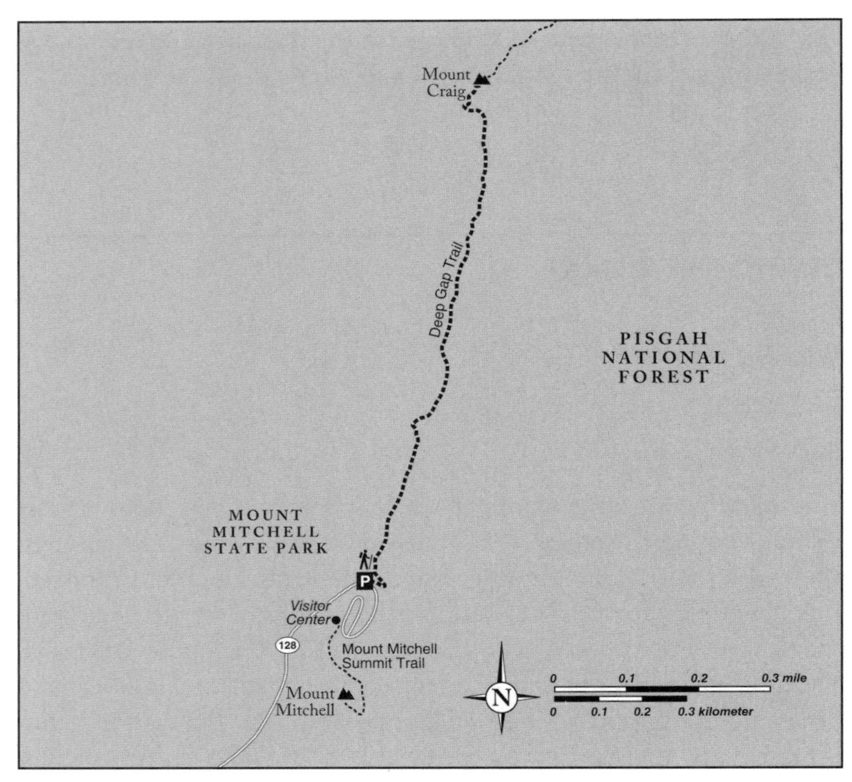

20. Mount Craig

gles. Along the way, keep a sharp lookout for the dense shrub thickets, spiny stems, and bright white leaf undersides of flowering raspberry (*Rubus idaeus*). From August into September, seek out the sweet ruby fruits, if you can spy them before the birds do! The ripest and tastiest berries are those plucked easily from the densely spiny stems.

This trail follows the ridgeline of the Black Mountains to Deep Gap, which connects to the Black Mountain Crest Trail that extends the length of the high-rising Black Mountains. This mountain range, composed of erosion-resistant volcanic and metamorphic rock, boasts six of the ten highest mountain peaks east of the Mississippi River, including Mount Craig, which at 6,647 ft is shorter than only Mount Mitchell. Your hike will take you 1.1 mi to the Mount Craig summit, where you will turn around and return to the parking area at Mount Mitchell.

20. MOUNT CRAIG

Highlights: Cool, dark forests reminiscent of New England; incredible views on a clear day from the Mount Craig summit

Natural Communities: Spruce-fir forest

Elevation: 6,684 ft (Mount Mitchell summit); 6,647 ft (Mount Craig summit)

Distance: 2.2 mi round-trip

Difficulty: Though there is little difference between the starting and ending elevations, this rugged trail has some steep descents to a saddle along a ridgeline, then a strenuous climb to the Mount Craig summit.

Directions: Mount Mitchell State Park is located in Yancey County. From the Blue Ridge Parkway junction in Asheville, go north to MP 355. Turn left at NC 128 and follow signs for Mount Mitchell State Park. Drive the winding road past the restaurant and campground toward the summit. Park next to the picnic area, just before the road bears right to continue to the summit trail parking area and visitor center (picnic area, concession stand, restrooms, and museum).

GPS: Lat. 35° 46′ 2.4″ N; Long. 82° 15′ 52.8″ W

Information: Mount Mitchell State Park, 2388 State Highway 128, Burnsville, NC 28714, (828) 675-4611

The spruce-fir forest you'll see along the trail is a boreal, or northern, forest at heart—a natural community more typical of northern New England or Canada. Our southern spruce-fir forests are dense, evergreen stands comprised mainly of Fraser fir (*Abies fraseri*) at these high elevations, but they also include red spruce (*Picea rubens*). Fraser fir, the most popular cut Christmas tree in the Southeast, is a southern Appalachian endemic; the species is found exclusively in this region. Fraser fir is replaced in spruce-fir forests to our north by the balsam fir (*Abies balsamea*). The other common coniferous species in our spruce-fir forests, red spruce, extends all the way to Canada.

Choose a dense section of forest and pause to hear the wind whisper through the trees. Breathe in the distinctive woodsy scent of this stand of Christmas trees that surrounds you. Occasionally you may hear the metallic "chip" of a slate-colored junco (*Junco hyemalis*) or see the flash of a tiny golden-crowned kinglet (*Regulus satrapa*), both birds more commonly found in New England forests. Even in these deepest patches of forest, however, human influence surrounds you, though hidden in the fragrant boughs of a recovering forest. Notice the rela-

Treeline and Latitude

If you have visited the White Mountains in New England, you may have observed that peaks exceeding 4,000 ft have treeless summits occupied by alpine tundra. Below 4,000 ft, these New England mountains have spruce-fir forest similar in many ways to that found in the southern Appalachians. Tundra replaces spruce-fir forest at higher elevations in New England because temperature decreases with increasing elevation, a phenomenon observed in mountains across the planet. As elevation increases, winters become too cold and too long for trees to exist, and spruce-fir forest gives way to open tundra of hardier plants. You might wonder why the Black Mountains of North Carolina, which tower over the numerous 4,000-footers in New England, have any trees at all. Shouldn't these high peaks in the Black Mountains also be clothed in treeless tundra? Why is Mount Mitchell, taller than Mount Washington in New Hampshire by almost 400 ft, forested to its summit?

The answer to these questions lies in a global pattern of increasing temperature with decreasing latitude. Underlying this pattern are a couple of simple physical principles. A given amount of sunlight is spread over a larger area at higher latitudes, so the sun's warming effect on the earth's

tively short and even stature of the Fraser firs. The similar ages of these trees indicate that they began life together as a single cohort. Look more closely and discover some remnant dead trees and many downed logs on the needle-covered forest floor.

After about one-third of a mile, you'll descend a rough-hewn stone staircase to the ridgeline saddle that separates Mount Mitchell from your destination, Mount Craig. You'll now encounter level walking for awhile, before you begin your steep ascent to the Mount Craig summit.

Imagine this as a pristine Fraser fir forest, as it looked before the arrival of the balsam woolly adelgid in the 1950s. These dark forests, with their thick stands of almost pure Fraser fir, were reduced to a stark landscape of whitened tree skeletons and sunny understory by a tiny insect native to Southeast Asia. Because the Black Mountains were among the first southern mountains hit by the adelgid as it swept from north to south, the traces of this devastation are now dis-

surface is reduced. Because sunlight passes through the earth's atmosphere at a low angle at higher latitudes, more of it is absorbed by the atmosphere before reaching the earth's surface. The bottom line is that it is simply colder, on average, the farther north or south one travels away from the equator.

If a hiker were to walk from south to north along the Appalachian Mountains while remaining at the same elevation, that hiker would experience a decrease in temperature, for every degree of latitude, equivalent to staying put in the south and climbing 350 ft higher. Given this equivalency of latitudinal and elevational effects on temperature, researchers have discovered that it is so much warmer in the Black Mountains (as compared to the White Mountains in New England) that Mount Mitchell, which is 6,683 ft, would need to be about 700 to 800 ft taller (about 7,400 to 7,500 ft) to be cool enough to discourage tree growth at its summit. Similar latitudinal trends are also found for the transition from northern hardwood forest to spruce-fir forest across the Appalachians. The southern Appalachians thus offer little in the way of suitable habitat for alpine plants, and those that do make their homes here must eke out an existence on rock outcrops and other such exposed habitats.

appearing as a new generation of Fraser fir seems to be thriving. Time will tell if some members of this new cohort of firs will have genetic resistance to the adelgid. In the meantime, the returning deep stillness of these recovering Fraser fir forests gives us hope for recovery and wonder at the incredible resilience of Mother Nature.

In the densest sections of forest, we encounter relatively few plant species. The canopy, mostly Fraser fir, is shared by occasional stems of red spruce, yellow birch (*Betula alleghaniensis*), and mountain ash (*Sorbus americana*). Scattered shrubs include witch hobble (*Viburnum lantanoides*), with its huge, rounded, and coarsely veined leaves; blueberries (*Vaccinium* spp.); Catawba rhododendron (*Rhododendron catawbiense*); and elderberry (*Sambucus pubens*). In the shaded understory, little grows aside from a carpet of splendid feather-moss (*Hylocomium splendens*), with occasional patches of mountain wood fern (*Dryopteris campyloptera*) and wood-shamrock (*Oxalis acetosella*). Pause for a few moments in one of the more open areas or along the edges, which are

Look for three cohorts of Fraser fir along the trail to Mount Craig: the downed logs of adelgid-killed trees from the 1950s, the current stand of even-aged trees that regenerated, and smaller saplings along the edges and in openings.

often overgrown with wildflowers. A visit here in early August reveals swaths of pink turtlehead (*Chelone lyonii*) and white snakeroot (*Ageratina altissima*) along the trail. This profuse floral display attracts an incredible variety of butterflies and bees.

At the summit of Mount Craig, be sure to stay on the marked trail to avoid crushing the fragile plants that cling to the bare rock here. Look for the round metal marker at the summit. If you're lucky enough to be atop Mount Craig on a clear day, look back toward Mount Mitchell and the crest of the Black Mountains to see where you've traveled.

You have the option, if you have more time to explore, to continue along the ridgeline of the Blacks to other peaks over 6,000 ft: Big Tom, Balsam Gap, Cattail Peak, and eventually (at mi 3.9) Deep Gap. At Deep Gap, there is a primitive shelter for backpackers, and the trail becomes the Black Mountain Crest Trail, which continues for a total of 12 mi to Bowlens Creek Road. However, our hike ends here, and you'll return the way you came to Mount Mitchell.

As you walk back toward the highest peak in the eastern United

Skeletal remains of Fraser fir can still be seen standing on the lower flanks of Mount Craig. The standing dead trees are slowly disappearing into the regenerating forest, but you can see from a distance that the new forest has a smooth appearance because the tree canopies are approximately the same height.

States, take time to appreciate the views or the swift-moving clouds, the whisper or howl of wind through the trees, and the dappled shade or the steady drum of cold rain on the forest floor, along with the ever-present woodsy scent of these spruce-fir forests. It's not hard at all to imagine yourself in the lovely, dark, and deep New England woods of Robert Frost.

21. BABEL TOWER AND WISEMAN'S VIEW

Why Are We Here?

Linville Gorge is wild, and the Babel Tower Trail can give you one of the best tastes of wilderness on the East Coast. It is steep and rugged,

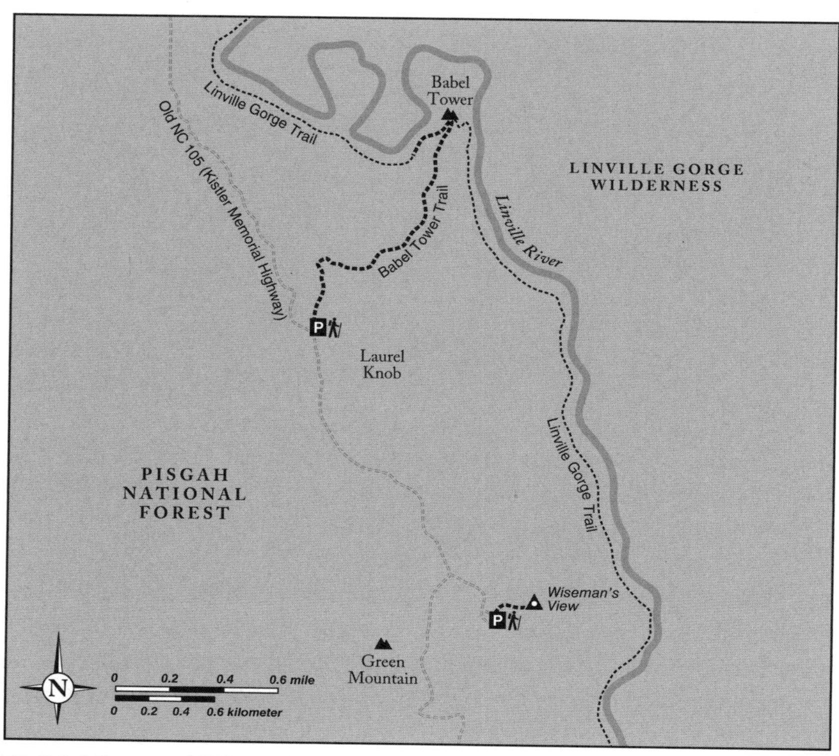

21. Babel Tower and Wiseman's View

and you'll clamber under and over fallen trees on your descent to the scenic Linville River. There are no trail blazes and few signs to guide you. You can get lost if you're not careful. Why come?

We recommend that you come to Linville Gorge Wilderness to replenish your soul. Humans, no matter how connected we are with technology and society, seem to need some contact with wilderness. It is a reminder of our ties and responsibilities to the natural world. As Aldo Leopold explains in *A Sand County Almanac* (1949), "In short, a land ethic changes the role of *Homo sapiens* from conqueror of the land-community to plain member and citizen of it." We must venture into the wilderness on occasion to renew our citizenship in the natural world. If you've never had the opportunity to experience a designated wilderness area before, the time is ripe for you to take this walk on the wild side.

As you hike, consider the long reach of human influence on the

21. BABEL TOWER AND WISEMAN'S VIEW

Highlights: Contemplate the meaning of wilderness and the long reach of human influence as you descend through this untamed federal wilderness to the wild and scenic Linville River.

Natural Communities: Montane pine forest and woodland, oak forest

Elevation: 3,800 ft (trailhead); 2,750 ft (Linville River)

Distance: 2.6 mi to Babel Tower and back; over 4 mi if you make the side trip to the Linville River and back

Difficulty: A rugged descent and strenuous climb back up to the rim of Linville Gorge; feels like more than twice the distance. This trail is unmarked and unmaintained; watch for downed trees and eroded gullies along the way and follow the trail closely. Sturdy boots are recommended.

Directions: Take US 221 south from the Blue Ridge Parkway to the intersection of NC 183 at the Linville Falls community. Turn left on NC 183 and drive 1 mi to its intersection with SR 1238 (Old NC 105, or the Kistler Memorial Highway); turn left there. This dirt road is rough and rutted in places. The Babel Tower trailhead is the fourth parking area on the left, with space for about a dozen cars (no facilities). Trailhead is marked with a single sign for Babel Tower that is not easy to see from the road.

GPS: Lat. 35° 55′ 11.9″ N; Long. 81° 55′ 17.8″ W (trailhead)

Lat. 35° 54′ 14.4″ N; Long. 81° 54′ 18″ W (Wiseman's View)

Information: US Forest Service, Grandfather Ranger District, 109 East Lawing Drive, Nebo, NC 28761, (828) 652-2144

montane pine forest and woodland (which gradually becomes an oak forest along the hike) you'll encounter along this trail. You'll learn that even with laws limiting human activity, continuing human influences are present long after the ink has dried on the papers designating a national wilderness. The forest you'll hike through is in decline because of several factors, including southern pine beetle (*Dendroctonus frontalis*) and suppression of wildfire. We'll discuss how such forests require fire to stay healthy and look for evidence of fire along our walk.

The Hike

The Linville Gorge Wilderness was one of the first designated by Congress in 1964 under the Wilderness Act. Now comprising 12,000 acres, this wilderness encompasses 12 mi along the Linville River. The Babel

Pines killed by southern pine beetle and charred stumps add to the wild character of the Babel Tower Trail in Linville Gorge Wilderness.

Tower Trail descends from the West Rim of the gorge toward the river, stopping at a column of rock that, if you climb carefully, provides great views of the gorge from within.

The first part of the hike descends through a forest of dying pines, and you'll pass by a primitive campsite on your right. Your first reaction may be a pang of disappointment; this is certainly not the cathedral of old-growth giants you'd find in a place like Joyce Kilmer Memorial Forest. Much of the canopy stands dead and dying around you. However, not all old-growth forests look the same, and the rugged topography and infertile soil prevent the establishment of a rich cove forest here. It's still old-growth—95% of Linville Gorge was never cut—but these forests are still very dynamic, owing to natural causes.

Montane pine forest and woodland communities are drought-adapted, clinging to mid-elevation dry ridges with thin, acidic soils, so it's no surprise that they are so prevalent in Linville Gorge, where such conditions abound. When you reach the first overlook, look carefully at the contours of the prominent ridges to see the dark evergreen trees hugging their crests and steep sides. The evergreens are three species of yellow pine: Table Mountain pine (*Pinus pungens*) at the highest elevations, with pitch pine (*P. rigida*) and Virginia pine (*P. virginiana*) at lower elevations.

This forest is what many ecologists would call pine-oak-heath, meaning that the canopy has a mix of drought-tolerant pines and oaks, while the understory is dominated by shrubs in the heath, or blueberry, family. According to our NatureServe classification system, this forest is transitioning from montane pine forest and woodland to oak forest; you are witnessing the prevailing pines being replaced by drought-resistant oaks.

As you hike, look carefully at the forest floor for seedlings. Why aren't pines regenerating? Quite simply, pine seedlings need full sunlight to grow, so they can't thrive in the deep shade cast by their parents. For pine regeneration to occur, something has to open the canopy to allow sunlight to penetrate to the forest floor. In addition, the thick layer of pine needles, called duff, that builds up under a mature pine stand prevents tiny, wind-borne pine seeds from becoming established. For successful germination and growth, pine seedlings need access to mineral soil. Dig through the needles to the mineral soil to assess the depth of the organic material, and also consider the amount of sunlight reaching the forest floor. The current conditions here prevent the replacement of pines by pines; tiny pine seedlings simply don't stand a chance.

In contrast to the yellow pines, oaks produce large and heavy fruits (acorns), which germinate and produce stout roots that can penetrate the thick layer of pine needles. In addition, oak seedlings will tolerate some shade. As you descend into Linville Gorge, you'll see species of oaks that prefer the dry end of the moisture spectrum, mainly chestnut oak (*Quercus montana*), with its large, scalloped leaves. If conditions remain as they are, these oaks will gradually take over the forest, outliving the pines that are established here now. Hike here again in twenty years, and you might find yourself standing in a dry oak forest.

Are these montane pine forests and woodlands, then, just a successional flash in the pan on their way to becoming an oak forest? Not necessarily. There is evidence that Linville Gorge and other places with dry, acidic ridges have maintained pines for extended periods. To learn more about the complexities of the natural cycle of fire and regeneration of montane pine forest and woodland communities, see the accompanying sidebar.

As you begin your descent, pause for a breather to take in the views and observe the forest around you. Interestingly, the fire in 2000 did not burn the canopy trees in your immediate vicinity. Indeed, most of the pines around you are more recently dead. These standing dead trees tell another piece of the story of Linville Gorge. They were killed by the native southern pine beetle *after* the 2000 fire. Researchers have tried to untangle the complicated relationship between fire and southern pine beetle in the southern Appalachians. It appears that both types of disturbance, while initially devastating to the existing canopy, interact and promote the continuation of pine forests in the mountains.

Pines killed by southern pine beetle often lose their bark in sheets, and if you stop and look closely at a beetle-killed tree that still has bark, you might see holes where female pine beetles have burrowed into the inner cambium. Though devastating to individual trees, southern pine beetle is a native insect, and researchers now think that infestations actually help perpetuate the montane pine forest and woodland. If low-level fires fail to clear oak saplings and remove the duff layer, a stand of beetle-killed pines on an exposed ridgeline is an open invitation for a lightning strike and a resulting hot fire, which create perfect conditions for pine regeneration. Fire adaptations of the two most prevalent pines in Linville Gorge, pitch pine and Table Mountain pine, are described in the sidebar "A Tale of Two Pines" in hike 22 (Table Rock).

The trail levels when you reach a junction with the Linville River Trail, which stretches 12 mi along the Wild and Scenic Linville River. The junction is not well-marked, and you may not see any signs. Just past this junction, you can climb up to the left on a rock outcrop to get your bearings and take a rest. The more prominent rock outcrop, Babel Tower, is still in front of you.

Enjoy the open views from the outcrop, then turn your attention to the trees in your immediate vicinity. There are fewer pines inside the gorge, but you'll notice an evergreen tree with short needles, Carolina hemlock (*Tsuga caroliniana*). The endemic Carolina hemlock is found only in a narrow geographic range in the southern Appalachians, and you can distinguish it from the more ubiquitous eastern hemlock (*T. canadensis*) by its bushy, bottlebrush branches with longer needles and larger mini-cones. Unfortunately, this species is susceptible to hemlock woolly adelgid (*Adelges tsugae*) like its more widespread relative, and it is rapidly disappearing from Linville Gorge.

You may also notice a small deciduous tree clinging onto some of the soil pockets around the outcrop. Sourwood (*Oxydendrum arboreum*) tolerates the dry, acidic soils of these rock outcrops and is our only tree representative in the blueberry family. In July, you'll see the white, urn-shaped flowers strung together and hanging down like sprays of seed pearls. These waxy flowers are enthusiastically visited by pollinators, especially honeybees. Sourwood honey is a southern Appalachian specialty, and it has a tangy flavor; look for it for sale at roadside stands near Linville Gorge. In fall, you'll easily recognize sourwood by its deep red, oval leaves.

The Babel Tower Trail continues only a short way farther, ending in a tower of rocks (Babel Tower) perched just above the river and affording you an excellent view of the interior of Linville Gorge. If you decide to negotiate your way forward and climb up to the tower overlook, use extreme care and keep in mind that you will have to return the way you came.

When you've explored the Babel Tower rock outcrop area as much as you would like, return to the junction with the Linville River Trail. Facing the rim of the gorge where you parked, turn right (north) if you want to make the short (less than half a mile) side trip to the river, considering that the climb out of the gorge will take you much longer than your descent. The trail continues downslope, but not as steeply. Walk the trail slowly and look to your right for a small spur trail that cuts down to the river. If the trail veers sharply left and away from the river, you've missed the turn.

Descend a short way through towering hemlock trees to the rocky walls of the Linville River. Each point along the Linville River has a different character; here, the river is constrained between tumbled rock walls, and a small waterfall rushes through a tight opening in water-smoothed rock before slowing and widening into a nice swimming hole below. The west side that you're standing on has smooth, broad rocks that invite you to stretch out. A good soak in the cold,

Fire on the Mountain

Why have montane pine forest and woodland natural communities persisted in places like Linville Gorge, when their ultimate fate seems to be replacement by drought-tolerant hardwoods? And why are they disappearing now? On the Babel Tower Trail, you may look for answers to these questions in the forest as you descend toward the Linville River. You'll soon realize that one important clue lies near your feet. Notice the blackened bases of the larger trees in this forest, living and dead. In November 2000, an unattended campfire spread and burned nearly 10,000 acres of the Linville Gorge Wilderness, including the forest in which you're standing.

Fire used to be a regular part of the landscape of Linville Gorge, but a long-standing policy of putting out all fires (publicly promoted by US Forest Service icon Smokey Bear) has eliminated most fires. Frequent fires in the past, started by lightning strikes on these exposed ridges, are thought to have burned away standing dead trees, low-growing shrubs, and the duff layer on the forest floor, allowing pines to become established and flourish. In the absence of fire, shade-tolerant oaks and other species have moved in, replacing the pines.

The historic fires were hot and devastating, leaping into the crowns of canopy pines and burning into the soil. When the smoke cleared, however,

rushing waters will rejuvenate your tired feet, and a quick rest and lunch will prepare you for the climb out of the gorge and back to the West Rim.

You may find yourself reluctant to leave this beautiful place, but sooner or later you'll need to tackle the return trip. Hike slowly and watch carefully for junctions with spur trails. After climbing through the large hemlocks next to the river, turn left to rejoin the Linville River Trail. When you reach the intersection with the Babel Tower Trail, turn right to climb up and out of the gorge, pausing again, if you wish, on the rock outcrop at the junction for a last inside view of the gorge.

As you hike out, you'll again have to duck under and clamber over fallen trees, beetle-killed pine, and adelgid-killed hemlock felled by winter ice storms and summer thunderstorms. What will this forest

the conditions were perfect for pine regeneration. The montane pine forest and woodland communities of Linville Gorge depended on fire for self-perpetuation. Considering fire scars on old trees and other evidence, researchers have determined that intense fires burned Linville Gorge in 1860 and 1915, with the last widespread fire in the 1950s. Until, that is, the 2000 fire.

Although the fire that burned in the winter of 2000 was widespread, it was relatively cool because of the season and the low density of the ground-level vegetation. In a few places where there were standing dead trees, the canopy burned, but for the most part, the fire stayed near the ground, burning mainly some organic material at the soil surface and shrubs in the heath family, such as mountain laurel (*Kalmia latifolia*) and gorge rhododendron (*Rhododendron minus*).

So Linville Gorge burned in 2000, but as you look around, you won't see many pine seedlings. The fire here was apparently not hot enough to regenerate pines. According to a number of studies that have reviewed the historic fire record, these montane pine forests and woodlands rely on a catastrophic fire every 75 years or so. As we've already discussed, the bare mineral soil that remains and ample sunlight are essential ingredients for germination of pine seed and successful establishment of seedlings.

look like in ten, twenty, or fifty years? A hot fire, fueled by the great amounts of dead wood, could open the canopy and reduce the soil duff layer enough to bring pine back into these forests. Without fire, dense thickets of oaks and shrubs will prevail. The only real certainty is that the forest will continue to change.

You may find yourself wishing the trail were better-maintained. However, no mechanized equipment is allowed in a wilderness area. The trees must lie where they fall or be cut by hand and dragged off the trail. Despite strict laws that limit human activity, Linville Gorge is still within the long reach of human influence. Through years of fire prevention and suppression, the cycle of fire, pine beetles, and pine regeneration has been interrupted. Humans have also introduced exotic invasive species like the hemlock woolly adelgid that profoundly influence the forest here by removing an important member of the natural community. [Note: As we go to press, debate is raging over the US Forest Service's proposal to reintroduce fire to Linville Gorge to restore pine and associated endangered species (mountain golden heather [*Hudsonia montana*] and Heller's blazing star [*Liatris helleri*]), as well as to reduce fuel loads to prevent out-of-control wildfires.]

Is there value in labeling Linville Gorge a wilderness, with so much human influence? As Leopold asserted in *A Sand County Almanac*, "All conservation of wildness is self-defeating, for to cherish we must see and fondle, and when enough have seen and fondled, there is no wilderness left to cherish." Can humans be visitors without intruding? Should we draw a line around Linville Gorge Wilderness and tell people not to cross? But that would ultimately be counterproductive; we must somehow see and value wild places, even if only a small subset of people (you now included!) have the ability to do so. We can appreciate the abstract concept of wilderness, but no concept, however lofty, can substitute for a direct experience, one that you can claim for yourself at Linville Gorge.

Side Trip: Wiseman's View

If you drive all the way to the Babel Tower trailhead on the Kistler Memorial Highway, you owe it to yourself to continue another mile, following the signs to turn left to Wiseman's View parking area. A handicapped-accessible trail (restrooms are also available) winds a short way through stunted montane pine forest and woodland to end at the rim of Linville Gorge, with breathtaking views of the gorge. Bring binoculars if you have them.

If you do not have the time or energy to hike the Babel Tower Trail, Wiseman's View makes for an excellent trip on its own because it will give you the best sense of the scale and wilderness experience from the western side of Linville Gorge, not to mention a bird's-eye view of the vegetation patterns.

Find a safe place to perch at any of the overlook areas and take in the view for a while. Here there can be no doubt why Linville Gorge has been called the Grand Canyon of the East. Look for turkey vultures (*Cathartes aura*) as well as hawks soaring on the rising air currents below you. The first overlook has a great view to the south (to your right) across the gorge to Table Rock. Look carefully for evidence of a wildfire that burned more than 4,000 acres in November 2013.

Table Mountain pine frames Wiseman's View looking north, which includes a foreground view of the craggy and rugged nature of these steep, dry montane pine forest and woodlands and the distant Grand Canyonesque view across Linville Gorge and down to the Linville River.

Go a bit farther down the trail, and you can make your way carefully through the trees to a perch with a view of Hawksbill and the entire gorge to the north.

Next, examine the vegetation patterns below and note how they vary with the topography. You'll see the darker green of evergreen pines hugging the dry ridges on the upper slopes, grading into other hardwood species in the ravines, which are more sheltered and moister. Some areas are steep enough that you'll notice barren ground and bare cliffs; here it is likely that plants cannot gain purchase on the steep slopes, or that existing vegetation has been torn out in a gully-washing summer thunderstorm. As your gaze travels farther down into the gorge, see if you can identify the transition from montane pine forest into montane oak forest. This is easier to do in the fall or winter, when you can use foliage (or lack thereof) to distinguish forest types. Remember, too, that Carolina hemlock will be tucked into the moist ravines and is evergreen as well, though typically a lighter shade of green than the pines. Also look for areas of dead, beetle-killed pine and possibly some dead hemlock in places near the Linville River.

As you scan different points in Linville Gorge, think about the environmental conditions of each site and how they help dictate the vegetation that grows there. Perched here on the edge of Linville Gorge, you may find yourself thinking like an ecologist, finding pattern and order in the landscape.

22. TABLE ROCK

Why Are We Here?

Table Rock is a popular hike, particularly in mid-October, when a clear day and a little effort to reach the top reward the hiker with a spectacular bird's-eye view of Linville Gorge, aflame with scarlet oak (*Quercus coccinea*) and brilliant sourwood (*Oxydendrum arboreum*). On a clear day you might see the Black Mountains across the gorge and off in the distance. Closer in, if you've brought binoculars, train them directly across the gorge and scan along the ridgetop to see hikers enjoying Wiseman's View (hike 21). You may also see rock climbers, laden with their heavy gear, preparing to climb to the summit of Table Rock.

As you begin your own climb to the summit, you'll see several pine species that are becoming less common in the southern Appalachians, and we'll discuss why they are disappearing. Several species thrive along this trail. Some coexist, and some are found only near the top or at the bottom. Clues along your path will help you see why.

The Hike

From the parking area's kiosk, head right to follow white blazes marking the Table Rock Trail. Sometimes you'll see a crew from NC Outward Bound timing a run to the summit. You'll want to proceed more slowly than these hard-core folks, to keep your oxygen levels steady and to observe the forest around you. Here at the base of Table Rock,

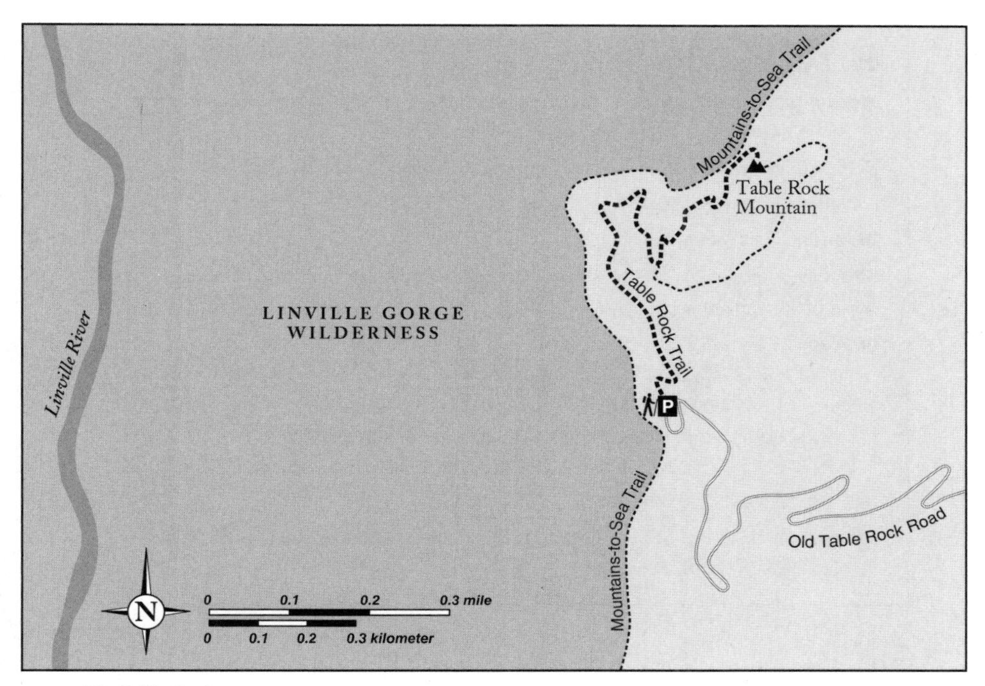

22. Table Rock

soil has accumulated in the slightly concave areas along the trail, allowing the development of a richer herbaceous understory. The first part of the hike is thus the only section that has any hint of soil richness; this is evident by the density and variety of wildflowers during the growing season. If you visit in July, you'll be treated to a variety of beebalms (*Monarda* spp.) in a rainbow of colors ranging from the familiar bright red to lavender, complemented by three-petaled, deep blue-purple spiderwort (*Tradescantia virginica*). Fall brings several species of asters with tiny, daisylike flowers.

You'll also see more tree species near the trailhead, including several oaks: northern red (*Quercus rubra*), white (*Q. alba*), and chestnut (*Q. montana*). However, Linville Gorge is an especially good place to study pines. All true pines worldwide are in the genus *Pinus*, but this genus is also divided into subgenera. In the southern Appalachians, all except eastern white pine are in the yellow pine subgenus, and these have coarse needles in bundles of twos or threes. Eastern white pine (*Pinus strobus*) is in the white pine subgenus, easily distinguished

22. TABLE ROCK

Highlights: The moderate climb to the top is just steep enough that you'll want to stop along the way to enjoy beautiful views of Linville Gorge.

Natural Communities: Montane pine forest and woodland, montane oak forest

Elevation: 3,200 ft (trailhead); 3,930 ft (summit)

Distance: 2.4 mi round-trip

Difficulty: A moderate and occasionally steep climb to the top, but with plenty of places to pause and enjoy the views

Directions: From US 221 in Marion, head north to Linville Falls and the intersection with NC 183. Turn right on NC 183, then right again on NC 181S. From there, go 3 mi and turn right on Gingercake Rd. (FR 210). When you come to the first fork, veer left to go through the Gingercake Acres development. Turn left on Table Rock Rd. to the Table Rock picnic area and parking lot (accessible pit toilets, accessible picnic area).

GPS: Lat. 35° 53′ 12″ N; Long. 81° 53′ 4.2″ W (trailhead)

Information: US Forest Service, Grandfather Ranger District, 109 East Lawing Drive, Nebo, NC 28761, (828) 652-2144

from the yellow pines by its feathery needles in bundles of five and its distinctive long, resinous, and tapering cones. Because this species produces a single whorled flush of branches each year, you can count the number of whorls and estimate the age of a tree. This pine is found in protected lower elevations throughout the southern Appalachians, which is the southern end of its range. As you explore the more exposed areas along the path to the summit, eastern white pine will disappear.

Abruptly the trail starts its moderately steep climb toward the summit along the side of the mountain, and here the soil turns dry and rocky. On the East Rim of Linville Gorge, the afternoon sun bakes Table Rock relentlessly, especially during the summer months. The heat and drought are relieved only by violent afternoon thunderstorms. The torrential downpours moisten the soil, but much of the rainfall quickly runs off the steep, rocky slopes and into the Linville River.

As you reach the base of the mountain and begin climbing steeply on switchbacks just a tenth of a mile from the parking area, you'll see a smaller trail branch off to the right around the base of Table Rock. This trail is primarily used by rock climbers. Be careful, then, as you walk around the summit, not to knock loose rocks over the edge and

The Chimneys, jutting up out of the East Rim of Linville Gorge, are an easy landmark to spot along the trail up to the Table Rock summit. A lone Table Mountain pine stands to the left, but the foreground shows that drought-tolerant oaks, such as chestnut oak, are slowly spreading into these drought-prone areas in the absence of fire.

potentially onto the heads of climbers, who are literally at the ends of their ropes!

As the trail narrows and becomes rockier, peek out some of the "windows" overlooking the gorge; these afford incredible views on a clear day. If you look back toward the parking area, you'll see a point jutting out from the edge of the gorge with two columns of rock at the end. This area is known as the Chimneys, named for those rock formations. Also notice how there is exposed bare rock at the lip of the gorge, while below it, the gorge slopes more gently toward the river. Long ago the Linville River sliced through the erosion-resistant quartzite at the edge and then cut more quickly through its growing canyon. It's amazing to consider how water can create a canyon as vast as Linville Gorge.

You'll be heading north now, and about a third of a mile from the parking area you'll need to switchback to the right to continue your

A Tale of Two Pines

At first glance, two of the species of yellow pine in Linville Gorge appear nearly identical. However, if the trees have tufts of needles poking out of the main trunk and their needles are grouped in clusters of three, you have discovered pitch pine (*Pinus rigida*). If the pine needles are in clusters of two and if rows of chunky cones encircle the upper branches, you have found Table Mountain pine (*P. pungens*). These species occur together in Linville Gorge, and both have special adaptations for fire.

The tufts of needles (and, if you look closely, buds) that are embedded in the bark of pitch pine give the species an advantage in the event of fire. If the fire is hot enough, the crown may be singed or even catch fire, leaving nothing but a charred trunk. Protected by the bark scales, the dormant buds burst into life following fire, allowing a scorched tree to regenerate its limbs. You'll notice several trees along the way whose trunks are covered with these epicormic sprouts (literally, sprouts upon the bark). Epicormic sprouting can also be a stress response of a tree exposed to very high light levels.

Table Mountain pine also responds to fire, but in an entirely different way. Each whorl of fat, closed, egg-shaped cones you see on the branches represents one year of reproductive output. You can often see over a decade's worth of cones together on some branches. They are waiting, sometimes for years, for the intense heat produced by a passing fire. These cones are serotinous, meaning late or late-developing. In this context, the

climb up the mountain, rather than going straight on the smaller Little Table Rock Trail. Notice as you climb that white pine has disappeared and has been replaced by gnarly, rough-barked yellow pines. Pause for a breather and examine these pines closely, then read the sidebar associated with this hike that shows you how to identify them.

Both Table Mountain pine (*Pinus pungens*) and pitch pine (*P. rigida*) are fire-adapted species, but why are they here? The high, rocky bluffs and ridges where these pines grow are vulnerable to lightning strikes from summer thunderstorms and the subsequent fires. In addition, these hardy trees are able to thrive in the thin, acidic soils found on rocky sites.

Shortly you'll head straight rather than going down and to the left,

*Pitch pine (*left*) is easy to identify, with small shoots sprouting along the trunk, while the hefty, armed cones of Table Mountain pine are arranged in rows of whorls around the branches. Both are different solutions to the problem of how to survive a fire-prone environment like Table Rock.*

term means that the cones are sealed shut with resin until heat softens the resin, allowing the cones to finally flex open their scales and release their seeds. If the tiny seeds were released each year, they would simply fall into the thick organic layer of pine needle duff on the forest floor, never to be seen again. However, a hot fire burns into the organic duff layer, and the heat subsequently opens these cones. The freed pine seed scatters into mineral soil with ample sunlight—perfect conditions for germination and successful establishment.

which would put you on the Mountains-to-Sea Trail, marked with round white blazes. Instead, you'll keep climbing. For the remainder of the hike, you'll see a few small paths that divert from the main trail (primarily ill-advised shortcuts that contribute to erosion), but continue to follow the main trail.

A deciduous species that might catch your eye as you hike is a small tree with coarse, irregularly toothed leaves. This is witch hazel (*Hamamelis virginiana*), especially attractive in mid- to late fall, when it might be flowering profusely. The bright yellow petals are narrow and delicate and jumbled together along the gnarled branches. Shortly after flowering, witch hazel produces woody fruits that are explosively dehiscent (they split open with force, which helps disperse their seed).

This odd seed-dispersal behavior might have advantages similar to those we discussed for Table Mountain pine. On Walden Pond in 1901, Henry David Thoreau brought witch hazel seed capsules into his cabin. In *Faith in a Seed* (1933) he wrote, "Three nights afterward, I heard at midnight a snapping sound and the fall of some small body on the floor from time to time. In the morning I found that it was produced by the witch hazel nuts on my desk springing open and casting their hard and stony seeds across the chamber. They were thus shooting their shining black seeds about the room for several days."

As you near the summit, stop to look at the structure of the vegetation around you. You may need to squint now if it's sunny, because you are no longer in a forest; the trees here are sparse and scattered. When you started, you were walking through a forest of white pine and several kinds of oaks. Now you'll only see an occasional pine on the exposed, rocky summit. Harsh winter weather may have something to do with this, but it's more likely that the thin soil prohibits development of the extensive root systems of trees. However, many shrubs in the blueberry family cover the mountainside, including highbush blueberry (*Vaccinium corymbosum*) and mountain laurel (*Kalmia latifolia*). They can tolerate the acidic, thin soil and hug the rocky surfaces, taking advantage of pockets of soil in cracks and crevices.

The summit of Table Rock has plenty of space. Even if it's crowded, you can find a private spot of your own. Take care, though, because there are steep drop-offs in all directions. During a recent visit, on one of those crystal-clear days in mid-October, we found what seemed like a solitary vista on Table Rock. Looking across the gorge for landmarks, we were startled to hear the clanking of metal against metal below us, where there was a sheer cliff. A few minutes later we heard voices as a pair of rock climbers popped over the ledge. They had taken the hard way to the top.

If it's a clear day, there's much to see from the summit. Looking north, the next summit over is Hawksbill (4,020 ft), a nice hike in its own right, though shorter and steeper, with fewer crowds and great views from the summit. The summit you see beyond Hawksbill is Sitting Bear, which occupies a ridge with Gingercake Mountain, the highest point in Linville Gorge Wilderness at 4,120 ft. Train your binoculars directly across the gorge and scan along the West Rim, looking for brightly dressed people taking in the gorge from Wiseman's View. If the day is clear enough, you should see a dark ridge of mountains well beyond the other side of the gorge; those are the Black Mountains, which boast 6 of the 10 peaks in the southern Appalachians over 6,000 ft. Looking to the south, you can see Lake James stretching out at the foot of the Blue Ridge Mountains, the recipient of the waters flowing from the Linville River.

Adjust your focus closer, inside the gorge, and see if you can discern forest patterns in the landscape. This is easiest to do in the fall and winter, when there is a clearer difference between deciduous and evergreen trees. Even in summer, though, you can distinguish the darker greens and spiky appearance of pines from the lighter greens and rounded crowns of deciduous trees. The fall color in Linville Gorge is spectacular, with the golden hues from chestnut oaks punctuated by deep red sourwood, red maple (*Acer rubrum*), and scarlet oak.

You'll see swaths of deciduous trees below you and on the gentle slopes lower in the gorge, which are more protected and have deeper soils (these are forests like the one where you started your hike, with white pine mixed with oaks and other hardwoods). Along the gullies and down near the Linville River, where soils stay moist, you can find evergreen Carolina hemlock (*Tsuga caroliniana*), which is only found in the southern Appalachians. Sadly, the swaths of dead trees are Carolina hemlock as well, mighty trees taken out by the tiny hemlock woolly adelgid (*Adelges tsugae*), an insect native to Southeast Asia against which our North American hemlock species have no defense. If this is the first time you've encountered the devastating effects of this tiny pest, please read the "Hemlock Woolly Adelgid" sidebar in hike 18 (Ramsey Cascades). Look next for the pine forests and heathy thickets that cling to the summits and rocky faces of the gorge, as well as areas so steep and rocky that they are either exposed or sparsely vegetated with sturdy shrubs in the blueberry family.

After enjoying the views and the landscape of the gorge, head back down the trail the way you came, remembering to keep to the main trail and to go left at the junction with the Mountains-to-Sea Trail. Now that you can distinguish the three species of pines along this trail, take special note of where they enter and leave the landscape. Watch for the disappearance of Table Mountain and pitch pines. Note also the first white pine you see on your descent. Where is it situated? Is it tall or stunted? What direction does the slope face? Where do you cross a line into continuous forest? By asking such questions and thinking through the answers, you'll be well on your way to interpreting this forested landscape.

NOTE: In November 2013, as this book went to press, a fire burned more than 4,000 acres, originating from the picnic area at Table Rock. While many lovers of the Linville Gorge Wilderness lamented the destruction of the forest, this fire might just be the disturbance needed to regenerate fire-adapted montane pine forest and woodland. As you hike to the summit of Table Rock, look for regenerating seedlings of both pitch and Table Mountain pine.

23. ROAN HIGH BLUFF AND ROAN GARDENS

Why Are We Here?

This excursion to Roan Gardens and Roan High Bluff is really two hikes in one: an easy 1-mi saunter on each of these trails will show you two distinctive high-elevation plant communities. Here there are no grass balds to be found, such as those on the Appalachian Trail hike just down the road and starting at Carver's Gap. Instead, your path will take you through a lushly sculpted shrubland and into a mist-shrouded spruce-fir forest.

Both hikes here are part of the Roan Massif, a large, mountainous mass with 5 distinctive peaks. Two of them, Roan High Knob and Roan High Bluffs, are on this western side of Carver's Gap, and both are shrouded in spruce-fir forest. Between them is a spectacular natural garden of Catawba rhododendron (*Rhododendron catawbiense*), which usually flowers in the third or fourth week of June, drawing visitors from all over the world.

We'll introduce you to two high-elevation communities of the southern Appalachians: shrub bald and spruce-fir forest. We have selected these hikes because they traverse some of the finest examples of both natural communities that exist. We'll also challenge you to

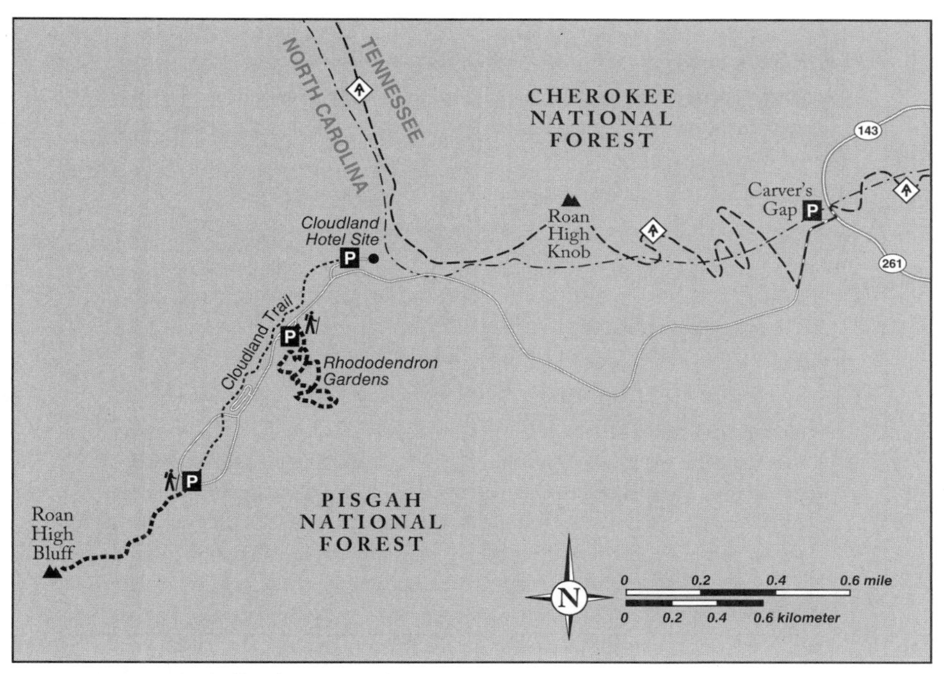

23. Roan High Bluff and Roan Gardens

consider the consequences of climate change for these rare and endangered natural communities.

The Hike (Rhododendron Gardens)

Begin your walk through the Rhododendron Gardens at the information cabin, where you can pick up a trail guide to the upper loop; this guide offers interpretive information for the numbered trail stops. The Rhododendron Gardens consist of two loops that together total about 1 mi. The upper loop, which is a third of a mile, is paved and accessible, while the lower loop is a natural-surface path.

You can wander through the gardens freely, alternating between taking in the views of the surrounding mountains and marveling at what appears to be a manicured, albeit natural, garden. The evergreen shrubs here certainly appear to be neatly arranged, with trim, rounded shapes that would look at home on the grounds of a grand

23. ROAN HIGH BLUFF AND ROAN GARDENS

Highlights: Spectacular natural gardens of Catawba rhododendron in mid- to late June, and a healthy high-elevation spruce-fir forest in the clouds. Two short hikes for the price of one!

Natural Communities: Shrub bald, spruce-fir forest, rocky summit

Elevation: 6,200 ft (Roan Gardens); 6,247 ft (Roan High Bluff observation platform)

Distance: Roan Gardens has two trail loops that total 1 mi; the upper paved loop is 0.3 mi. The short section of the Cloudland Trail featured here is 1 mi round-trip; the entire trail is 1.2 mi round-trip.

Difficulty: Easy and level walking in Roan Gardens. The upper loop of Roan Gardens is paved and accessible. The short section of the Cloudland Trail is also an easy hike.

Directions: From Bakersville, NC, drive north on NC 261 for 13 mi to the NC/TN line at Carver's Gap (on the TN side, the road is TN 143). Turn left on the entrance road (open from late May to October, but call ahead to be sure, if you're in the "shoulder" season), and pay a small entrance fee at the US Forest Service kiosk. Parking for the Roan Gardens is just over a mile on the left, where the paved road turns to gravel. To hike all of the Cloudland Trail, park at the Cloudland Hotel Site, just before the Rhododendron Gardens. To walk the short section to the Roan High Bluff, which is described here, continue on the gravel road to the loop and park along the road to access the trail (accessible restrooms and picnic area at Roan Gardens).

GPS: Lat. 36° 5′ 47.4″ N; Long. 82° 8′ 26.4″ W (Roan Gardens)

Lat. 36° 5′ 35.4″ N; Long. 82° 8′ 43.4″ W (Roan High Bluff)

Information: Appalachian Ranger District, Pisgah National Forest, P.O. Box 128, Burnsville, NC 28714, (828) 682-6146

estate. Yet Mother Nature is the only gardener at work here, using winter ice and chilling winds to prune errant branches.

The star of the show, Catawba rhododendron flowers at these high elevations during the latter half of June. The Roan Mountain Rhododendron Festival, now taking place down the mountain on the Tennessee side, at Roan Mountain State Park, is held on the third weekend in June each year. This area has been jointly managed by Cherokee National Forest on the Tennessee side and Pisgah National Forest on the North Carolina side since 1941, when it was acquired by the US Forest Service. The natural gardens cover hundreds of acres and are protected, though as recently as the 1930s many Catawba rhododendrons were dug and sold at nurseries.

You'll see as you wander the trails that while the gardens are

The ecotone or boundary between spruce-fir forest and shrub bald (which is nearly 100% Catawba rhododendron) is very distinct at Roan Gardens.

clearly shrub-dominated, there are some red spruce (*Picea rubens*) and Fraser fir (*Abies fraseri*) encroaching on the edges. These trees have the potential to shade out the shrubs that currently dominate. For now, however, no management other than what is provided by the harsh climate is needed to keep these gardens looking their finest.

Once you have completed your tour of the Rhododendron Gardens, drive to the end of the road for something completely different!

The Hike (Cloudland Trail)

The hike featured here is a short, half-mile segment of the Cloudland Trail, which ends at Roan High Bluff. The trail to the overlook is marked with gold rectangular blazes, but it is easy to follow the fir-needle-carpeted path through a dense spruce-fir forest that smells like Christmas.

If you have hiked in spruce-fir forest before, you might be struck by the healthy look of this forest. This is especially true if you compare the spruce-fir forest here with similar forests at Mount Mitchell and especially Clingman's Dome, the latter in Great Smoky Mountains National Park. The forest at Mount Mitchell is recovering from infestation of the balsam woolly adelgid (*Adelges piceae*) in the 1950s, with stands of young, similar-sized trees replacing the bare skeletons of the former forest. Clingman's Dome, which did not see the adelgid until the 1970s, still looks ravaged, with a stark landscape of whitened tree skeletons and open, sunny gaps.

Here, if you look carefully, you will also see the bare trunks of dead firs as evidence of the adelgid's depredations, first detected on Roan Mountain in 1962. However, the skeletons are few in number and hidden by the lush new growth of young fir, as well as taller specimens. Why does this spruce-fir forest seem less affected by this exotic insect from Southeast Asia? Why does this forest have trees of all ages, rather than a single cohort that germinated when the older trees succumbed to the adelgid?

Perhaps surprisingly, the answer might be tied to human influence at Roan prior to the infestation of the adelgid. Extensive logging until the 1930s meant that this forest stand was younger than the forests on Mount Mitchell and Clingman's Dome when the adelgid arrived in the early 1960s. Researchers speculate that many of the younger trees survived the adelgid attack that devastated more mature stands of trees at other locations.

Enough mature trees fell in this forest, however, to offer opportunities for shade-intolerant fir seedlings to find a foothold, sometimes on top of their fallen brethren. Fir is considered an early successional species that grows quickly into gaps with ample sunlight. Here, above 6,000 ft, fir is more abundant than red spruce, and the mix is about 80% Fraser fir to 20% red spruce. Today's forest is healthy enough that, in early fall, the US Forest Service allows limited collection of fir seed to supply Christmas tree farmers in the surrounding area.

At a half mile, the trail descends to a wooden overlook platform. You have reached Roan High Bluff, the westernmost peak of the Roan Massif, at 6,267 ft. The steepness of the north-facing slope here means that any trees that survive are short and stunted, and many areas are completely treeless. Here the community ceases to be a forest and is instead a rocky summit. Few trees means that more sunlight reaches the rocky surface, providing microclimates for some federally endan-

On a misty day, you will feel as though you are on the edge of the world on the north-facing, 6,267-ft Roan High Knob, a prime example of a rocky summit natural community. While flag-form fir populates the top of the knob and a few of the deeper pockets of soil, rare plant species keep a tentative grasp on life in the rocky crevices.

Climate Change on Sky Islands

The spruce-fir forests protected at Roan and other high mountains of the southern Appalachians represent one of the most endangered natural communities in the world. These "sky islands" of forest appear only on mountain peaks above 5,500 ft. On the Roan Massif, the US Forest Service protects 6 federally listed endangered species: cliff avens (*Geum radiatum*), Roan Mountain bluet (*Houstonia montana*), Blue Ridge goldenrod (*Solidago spithamaea*), rock gnome lichen (*Gymnoderma lineare*), Carolina northern flying squirrel (*Glaucomys sabrinus coloratus*), and spruce-fir moss spider (*Microhexura montivaga*). All depend on the cool, moist environments provided courtesy of the extreme climate.

Southern Appalachian spruce-fir forests have contended with a number of challenges over the course of their existence. About 6,000 years ago, there was a period of marked global warming called the hypsithermal. During this period, spruce-fir forests migrated to higher elevations. Lower mountains that had no cool-weather refuge lost this natural community altogether. Acidic precipitation and the balsam woolly adelgid (*Adelges piceae*) represent more recent threats, with the adelgid accounting for 95% of the observed loss of Fraser fir (*Abies fraseri*). Current models predict that global climate will continue to warm rapidly, an additional threat to these forest systems. Spruce-fir forests already grow on the tops of the mountains, so warming climate may eliminate them. Reduced precipitation predicted by some climate models could also make these forests vulnerable to catastrophic fire, to which red spruce (*Picea rubens*) and Fraser fir are not well-adapted.

Climate change is also a foreboding challenge for rare species of high elevations in the southern Appalachians, because a warming climate will push many of these species off their remaining mountaintop habitats (the so-called sky islands) forever. For the disjunct species, such as three-toothed cinquefoil (*Sibbaldia tridentata*), deerhair bulrush (*Trichophorum cespitosum*), highland rush (*Juncus trifidus*), and Appalachian firmoss (*Huperzia appressa*), other populations far to the north would presumably remain. For the endemic species (cliff avens and Cain's reed-grass [*Calamagrostis cainii*], for example), loss of their southern Appalachian mountaintop refuges would spell almost certain extinction. Like the woolly mammoth (*Mammuthus primigenius*), these species would join the legions of other species that could not sustain the dual postglacial onslaughts of climate change and burgeoning human populations.

Fortunately, most of the acreage of southern Appalachian spruce-fir forest is owned by the federal government already, but simply protecting it from harvest may not be enough to preserve it. However, it may be premature to give these forests up for lost. Scientists believe that temperatures during the hypsithermal were much warmer than today, and yet the spruce-fir forests we see today are descendants of those that found refuge on the highest peaks. And so there is hope that our great-grandchildren will one day walk among these trees' shady boughs, taking in their wintery scent.

gered shrubby and herbaceous plants, such as spreading avens (*Geum radiatum*), which has large, rounded leaves and yellow flowers that bloom in late July, and the Roan Mountain bluet (*Houstonia montana*), an inconspicuous plant with four-petaled lavender flowers.

If the clouds have lifted enough for you to catch the view, you can see the community of Buladean and the Unaka Mountains. More often than not, however, you'll feel as though you are perched on the edge of the world, with nothing below your feet but clouds.

24. GRASSY RIDGE

Why Are We Here?

The Roan Massif comprises five separate peaks: Roan High Knob, Roan High Bluff, Round Bald, Jane Bald, and Grassy Ridge. It sits on the North Carolina/Tennessee line and presents a fascinating array of high-elevation ecosystems with clearly visible boundaries. The dense forests of spruce and fir give way abruptly to shrub balds (also known as heath balds) of Catawba rhododendron (*Rhododendron catawbiense*) that bring visitors from near and far to view their floral display in late June (see hike 23 [Roan High Bluff and Roan Gardens]). Other peaks are

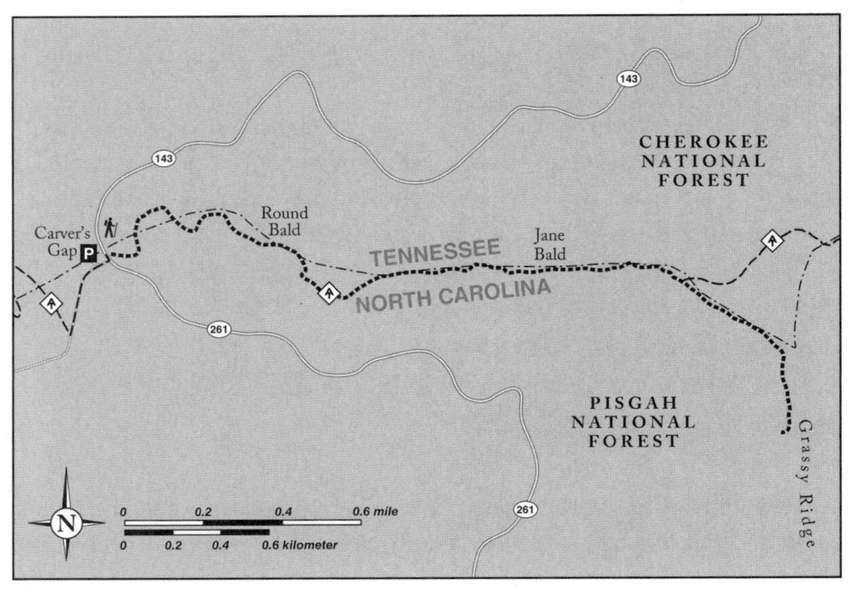

24. Grassy Ridge

24. GRASSY RIDGE

Highlights: Some of the most spectacular views on the Appalachian Trail; surprisingly abrupt transitions between forests, grasslands, and natural rhododendron gardens

Natural Communities: Spruce-fir forest, shrub bald, grass bald

Elevation: 5,512 ft (Carver's Gap); 5,826 ft (Round Bald summit); 5,807 ft (Jane Bald summit); 6,158 ft (US Forest Service marker on Grassy Ridge)

Distance: 2.5 mi one-way; 5 mi out-and-back; options to shorten the hike from any of the closer peaks, Round Bald or Jane Bald

Difficulty: Mostly easy walking over the rolling peaks of the Roan Highlands, with moderate climbs to Round Bald and Grassy Ridge

Directions: From Bakersville, NC, drive north on NC 261 for 13 mi to the NC/TN line at Carver's Gap (on the TN side, the road is TN 143) and park in the large parking area (pit toilets). Cross the road to start your hike.

GPS: Lat. 36° 6′ 22.7″ N; Long. 82° 6′ 37″ W (Carver's Gap trailhead)

Information: Appalachian Ranger District, Pisgah National Forest, P.O. Box 128, Burnsville, NC 28714, (828) 682-6146

dominated entirely by grass species, giving way to shrubs and northern hardwood forests on the saddles between them. The question here, then, is why do we see forest in one place and no trees whatsoever elsewhere? The answers aren't clear—which makes this question all the more interesting.

Beyond the abrupt transitions in vegetation, the views on this trail on a clear day are spectacular. This part of the Appalachian Trail from Carver's Gap to Grassy Ridge is said by many to be the most scenic section of the trail, which is more than 2,000 mi long. The most popular time to visit the Roan Highlands is the last two weeks of June, when Catawba rhododendrons set the mountainside aflame with their deep pink to purple flowers.

Through-hikers on the Appalachian Trail aren't the only ones who have enjoyed the views at Roan; the massif also attracted early botanists such as André Michaux and Asa Gray. More than 300 species of plants, 33 of them identified as rare, have been documented in this relatively small area. Roan's status as a unique and biologically diverse natural area has led to long-term conservation of more than 15,000 acres, held and managed by the Pisgah and Cherokee National Forests, the Nature Conservancy, and the Southern Appalachian Highlands Conservancy.

The Hike

From the parking area at Carver's Gap, you'll cross NC 261 and walk through a fence stile to start an easy climb along the Appalachian Trail. The trail follows a series of switchbacks and suddenly enters a cool, dark forest of yellow birch (*Betula alleghaniensis*) and red spruce (*Picea rubens*), with Fraser fir (*Abies fraseri*) increasingly evident as you climb. This is the only forest you will pass through on this hike, though you will see plenty of forested areas on the surrounding mountains.

Emerge from the forest into sunshine (or rain, as the case may be) after hiking through this short stretch of forest and look around you. The elevation has not changed significantly, yet you're now walking through a transition area of trees, shrubs, and herbaceous plants, none of them dominant, before emerging into the open meadows that dominate Round Bald. If you have visited other grass bald hikes in this book, you may be inclined to say, "OK, another grass bald—I wonder how they are going to explain this one?" To this we can reply that, at Round Bald, we have evidence from an experimental study that may shed some light on the absence of forest on the summit. In this case the experiment involved the planting of Fraser fir, which is what you'd expect to be present at this elevation. To learn more about this study and its inconclusive results, please read the sidebar for this hike.

In addition to this hike's sidebar, you may also wish to read "Grass Bald Origins and Future" in hike 13 (Hooper Bald) and the "When a Grass Bald Isn't" sidebar in hike 29 (Mount Rogers).

As you crest Round Bald, you'll pass the summit marker at 5,826 ft. Any of the summits you'll cross on this trail make for a fine turnaround point, so if you only have time for a short hike, pause to enjoy the panoramic views before heading back to Carver's Gap. If the day is fine and the meadows are calling, continue hiking, descending to Engine Gap, between Round and Jane Balds.

Grass balds in the southern part of the southern Appalachians seem to be dominated by European grasses; however, Roan boasts some rare species found only in these unique, high-elevation environments. In late June you may see the nodding, deep-red-speckled trumpets of Gray's lily (*Lilium grayii*) in the open areas. This lily is named after Harvard botanist Dr. Asa Gray, who discovered the plant at Roan in 1841. These rare beauties grow only on open, high-elevation mountains in North Carolina, Virginia, and Tennessee.

The trail begins to climb gently again as you walk toward the summit of Jane Bald. From August into September, you'll see the tall, flat-topped flower clusters of filmy angelica (*Angelica triquinata*) standing at full attention above the grasses like giant, greenish versions of a related and familiar wildflower, Queen Anne's lace (*Dauca carota*). Chances are, the flowers will be swarming with insect pollinators and you may hear them buzzing as you approach. If you pause to watch them for a few minutes, you'll notice that they are unapologetically drunk. Many are so blitzed that they can't fly; researchers have, in the name of science, knocked them to the ground and seen them lie there, stupefied, for up to 3 minutes. Filmy angelica, despite having open flowers, has abundant stores of nectar that contain some kind of narcotic irresistible to its slap-happy pollinators. However, the actual substance that leaves bees stupefied remains a mystery.

Gardening for Ecological Explanations

As you climb toward the open summit of Round Bald, which is less than a mile from Carver's Gap, look to your left for a patch of spruce-fir forest. These trees were planted in 1928 in a classic garden experiment to answer a basic ecological question: is the environment here unfavorable for tree growth? If the answer is yes, we would expect that these trees would not have survived. In other words, something about the local environmental conditions prohibits their growth, which is why they were not here to begin with.

If the trees do survive, and they do, we might claim that some other factor, rather than the environment, previously limited their occurrence here. The nearest populations may not be close enough or the prevailing winds might not blow in the right direction to disperse seeds successfully from nearby populations to these treeless summits. Thus they don't occur here because they haven't had the opportunity to disperse here. There may be another explanation, such as a past disturbance that eliminated these trees.

Actually, we can say none of these things with certainty. Nature is always more complicated than we might like. The trees persist. However, if you walk over to the stand, you will notice that, unlike the stands along the Cloudland Trail leading into the spruce-fir forest on Roan High Knob, this stand is in a kind of suspended animation. The trees grow—slowly—yet there are few seedlings at their feet. Perhaps the former absence of trees on Round Bald is a combined effect of missed opportunity (biogeography) and less-than-ideal growing conditions (environment).

As you climb toward the summit of Jane Bald, you'll notice more shrubs around you. Each of the balds along this trail has a different character, with Round being the most open and the slopes below Grassy Bald being the most closed. Among the Catawba rhododendron you'll notice a shrub with toothed leaves widest toward the tip. This is green alder (*Alnus viridis* var. *crispa*; the "var." in this name refers to a named variety of green alder), which forms dense thickets in a very few high-elevation bald areas, with the rest of its populations found in Pennsylvania and farther north. This wide geographic separation, or disjunction, makes this a rare shrub community at Roan, though green alder is abundant here.

The summit of Round Bald, as seen from Jane Bald, is the most open of the three grass balds on this hike. Notice the encroaching shrubs and trees on the edges, which are kept back by hand-cutting, mowing, and grazing by goats. Beyond Round Bald, you can easily see Roan High Bluff, which is cloaked in dark spruce-fir forest.

We can speculate that alder-dominated communities were more widespread in the southern Appalachians during cooler times, and that these populations are remnants surviving at high elevations. Current management dictates, however, that the ecological goal for Roan is to maintain the grassy bald community, so green alder, along with other shrubs, is under constant attack from mowing, lopping, and sometimes grazing. There is some disagreement among land managers about cutting down the alder to favor grasses and herbs, because the alder-dominated community is also rare.

Before cresting the summit of Jane Bald (elevation 5,807 ft), you'll walk up and over a rock outcrop area with very distinct horizontal fissures. Walk to the untrampled edges of this rock and look carefully in the crevices for three-toothed cinquefoil (*Sibbaldia tridentata*), which is abundant at Roan Mountain but listed as a state endangered species. This plant, like green alder, is more abundant to the north. Indeed, three-toothed cinquefoil occurs as far north as Newfoundland and Greenland. Further reading on glacial relict plants may be found in the "Ice Age Relicts" sidebar in hike 11 (Satulah Mountain).

Continue down the back side of Jane Bald, and before long you'll reach a trail junction for the Grassy Ridge spur trail. The spur trail goes straight ahead, while the Appalachian Trail branches to the left. You'll continue straight and begin a moderate climb up Grassy Ridge. Shortly after the trail junction, you'll enter a tunnel of dense shrubs composed primarily of Catawba rhododendron and green alder, plus some blueberry (*Vaccinium* spp.). You are smack in the middle of a heath bald. The environmental conditions underlying the heath bald here are different from those in other parts of the southern Appalachians, where we think of heath balds as communities clinging to the thin, acidic soils that flank steep-sided mountains. In such environments, these shrubs are among the only species that can survive. Here, though, the slopes are gentle, but the shrubs are clearly outcompeting herbaceous species.

This section of trail was described eloquently by Dr. Elisha Mitchell, who in his 1836 personal journal named Roan his favorite mountain. "The top of Roan may be described as a vast meadow without a tree to obstruct this prospect, where a person may gallop his horse for a mile or two with Carolina at his feet on one side and Tennessee on the other, and a green ocean of mountains rising in tremendous billows immediately around him."

You will emerge out of the heath bald into an open, grass bald along Grassy Ridge again, and when you do, pause to catch your breath from the climb and look at the panoramic views. Dr. Mitchell's green ocean of mountains stretches out before you, if the day is clear, but today Mitchell would not have a mile or two of open meadows to gallop his horse. Everywhere you look, patches of shrubs dot the formerly open areas. The shrubs are clearly taking over.

Historical evidence, such as Mitchell's journal, supports the notion that these grass balds have become shrubby since Mitchell's time, and without human intervention, they would close in still more rapidly. Today, Grassy Ridge and much of Roan Mountain is kept open through active management by the US Forest Service, the Southern Appalachian Highlands Conservancy, and other conservation organizations. Annual "mow-offs" use power equipment to cut back the shrubs, and some areas are grazed. Big Yellow Mountain, a nearby Nature Conservancy preserve, has been kept open through continuous cattle grazing by the Avery family for over 100 years.

Grazing, rather than burning or mowing, the grassy balds as a management technique is more in line with current thinking that these balds were once treeless alpine tundra environments during the most recent Ice Age, about 10,000 years ago. Once global temperatures warmed, these areas stayed open, grazed by Pleistocene megafauna such as ground sloths (*Megalonyx jeffersonii*) and woolly mammoths (*Mammuthus primigenius*), which were eventually replaced by elk (*Cervis canadensis*) and bison (*Bison bison*) and, more recently, by the livestock of early settlers. Now that the mountaintops are devoid of hoofed grazers and their ilk, these areas are slowly being invaded by shrubs. Further discussion of this topic may be found in "Grass Bald Origins and Future" in hike 13 (Hooper Bald).

If you visit Roan during the summertime, you may pass by a mailbox on a pole marked BAAATANY, with brochures inside detailing one management method for maintaining the balds as open areas of grasses and herbs. Each summer since 2008, Jamey Donaldson has taken his herd of goats up to graze sections of the balds along the Appalachian Trail. The goats are especially fond of blackberry, and the program organizers feel that using grazers to maintain the balds is the most historically accurate method of keeping them open with disturbance.

During your climb up Grassy Ridge, you will cross the 6,000-ft elevation mark, and at 2.6 mi, you'll come to a US Forest Service memorial dedicated to Cornelious Rex Peake, a local farmer who advocated for the conservation of the Roan Highlands. His nearby farm was the highest cultivated part of the highlands, and he was instrumental in the mid-20th century in getting protection for the Roan Highlands. Look east for views of Grandfather Mountain and south for views of the Black Mountains.

The rock outcrop here makes for a great rest stop, and you've reached the end of the hike. Open views like this are rare in the forested southern Appalachians, so take time to look behind you and trace your path back to the Round Bald summit, beyond which lies Carver's Gap. There is something immensely satisfying about seeing your overland route and the miles you've covered on your own two feet.

When you can tear yourself away from the view (or if driving rain hurries you along), return on the spur trail and keep straight ahead as you reach the intersection with the Appalachian trail, heading south toward your car rather than turning right and heading toward Maine. As you walk, notice the character of each of the balds as you see them in front of you: the shrubs creeping up the sides of Jane Bald, the open summit of Round Bald, and the forested slopes of Roan High Bluff beyond Carver's Gap. We hope you now have a better understanding of the natural phenomena and human intervention that maintain this complex mosaic pattern of vegetation.

25. GRANDFATHER MOUNTAIN

Why Are We Here?

The Grandfather Mountain Trail perfectly captures this mountain of extremes. Its geographic position farther north than the taller peaks in the Black Mountains (such as Mount Mitchell) and its precarious perch on the edge of the Blue Ridge Escarpment give it a rugged, alpine feeling. In fact, in 1794, French botanist and explorer André Michaux was said to have jumped up and down on the summit of Calloway Peak and burst into a chorus of *La Marseillaise*, so convinced was he that he had discovered the highest summit in eastern North America. In reality, Calloway Peak is just below the 6,000-ft mark at 5,954 ft. But the steepness of Grandfather and its north-facing cliffs make this a place of extremes, which in turn makes life difficult for those who live here. You'll explore these harsh edges and learn how plants and animals survive in such a challenging environment.

The second part of the hike descends to NC 105 along the Profile Trail. Here you'll leave the extremes behind as you enter a northern hardwood forest, which transitions into a beautiful rich cove forest as you descend. Along the way, you will see how species are specially adapted for the specific habitats where they are found.

The Hike

Begin your hike across the parking lot from the steps that lead most visitors to the Mile-High Swinging Bridge (if this is your first visit to

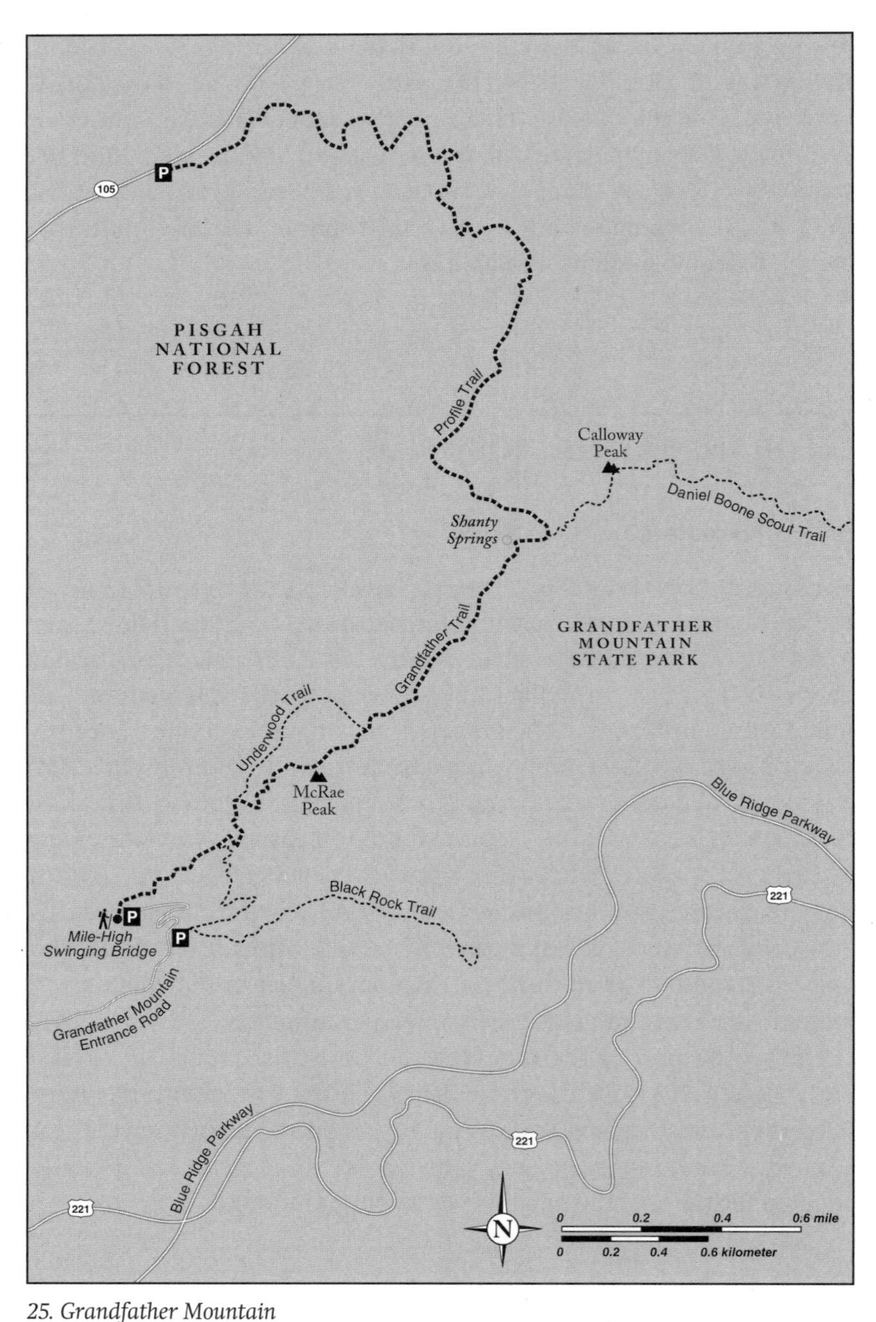

25. Grandfather Mountain

25. GRANDFATHER MOUNTAIN

Highlights: An extreme trail featuring extreme habitats and the species that inhabit them, this hike will challenge your physical abilities while enabling you to see how plants and animals eke out a living in challenging environments.

Natural Communities: Rocky summit, spruce-fir forest, northern hardwood forest, cove forest

Elevation: 5,230 ft (Grandfather Trail trailhead); 5,844 ft (McRae Peak); 3,880 ft (Profile Trail trailhead)

Distance: 1.9 mi from the Mile-High Swinging Bridge parking lot to the junction with the Profile Trail, then 3.1 mi to the parking area on NC 105, for a total one-way trip of 5 mi. Another option is an out-and-back hike to Calloway Peak from either the Profile trailhead or the Mile-High Swinging Bridge parking lot.

Difficulty: Very difficult, including sections where you must use ladders and cables to climb bare rock faces. There is an alternate section, the Underwood Trail, which is recommended in poor weather or for anyone who is uncomfortable with heights. Either way, it's a strenuous hike. Sturdy boots are a must.

Directions: The best way to set up this hike is to park one vehicle at the bottom of the Profile Trail, on NC 105, 0.7 mi north from the junction of NC 105/NC 181. Then drive a second vehicle south on NC 105 and turn left onto US 221. Turn left on the Grandfather Mountain Entrance Rd. Continue following the road to the parking area next to the Mile-High Swinging Bridge. The trailhead starts across the parking lot from the swinging bridge stairway (entrance fee, restrooms, picnic areas, restaurant, exhibits). On busy weekends, you'll be asked to park at the hiker's parking lot just below the mountain, adding a pleasant 0.4-mi (one-way) walk up to the Swinging Bridge parking area.

GPS: Lat. 36° 7′ 19.3″ N; Long. 81° 49′ 48.3″ W (Profile Trail trailhead on NC 105)

Lat. 36° 5′ 45″ N; Long. 81° 49′ 56″ W (Mile-High Swinging Bridge parking lot)

Information: Grandfather Mountain State Park, P.O. Box 9, Linville, NC 28646, (828) 963-9522

Grandfather Mountain, take time to walk across the chasm). A sign appropriately warns that the Grandfather Trail is strenuous and that sturdy footwear is required. Follow the blue blazes along a rugged, rocky trail that is punctuated by deep pink Catawba rhododendron (*Rhododendron catawbiense*) in May.

Everything about this section of trail screams extreme. Wind speeds of 190 mph have been recorded at the weather station, and the swinging bridge is frequently closed due to high winds, driving rain, and ice. Be sure to check the weather forecast before you set out on your adventure, and be prepared for weather contingencies.

You'll climb the first half mile through mixed red spruce (*Picea rubens*), Fraser fir (*Abies fraseri*), and northern hardwoods. The canopy becomes shorter as you get closer to the exposed ridgeline, and you'll see high-elevation shrubs such as witch hobble (*Viburnum lantanoides*), mountain ash (*Sorbus americana*), and Catawba rhododendron reaching into the canopy.

The ridgeline you are following runs north/northeast to south/southwest. In the lower gaps, you may notice west- to northwest-facing openings that are subject to the full fury of the prevailing westerly winds. Here you'll discover cliffs of scree (loose rocks) and patches of low-growing blackberry; in these tight places, the wind funnels through and prevents vegetation from establishing. The few trees you see are stunted.

The plant communities along this section of trail alternate between gnarled spruce-fir forest and a high-elevation rocky summit. In both community types, the vegetation shows the effects of the harsh environment in which it lives. Trees grow in flag form, with branches only growing on the side away from the prevailing westerly winds. The branches that grow toward the exposed side are pruned off by strong winds and winter ice. Shrubs and perennials hug the jagged surfaces of the rocks and hunker down between crevices. Acute weather conditions dictate and shape these communities.

Look closely and notice that the trees that delineate a spruce-fir forest can only exist where there is enough soil for Fraser fir and red spruce to gain a foothold. You might see an occasional fire cherry (*Prunus pensylvanica*) here as well. Fir is most prevalent at the highest elevations, but watch for more and more red spruce when you descend the Profile Trail. These two evergreen trees aren't always easy to tell apart at first glance. Fir needles are flat and soft; spruce needles are round (you can roll them between your fingers) and pointy.

After about a half mile, you will come to a junction with the Underwood Trail, marked with yellow blazes. The Underwood Trail provides a more moderate choice for hikers not wanting to undertake the cables, ladders, and exposed cliffs on the next section of the Grandfather Trail. Additionally, this trail is a better choice for everyone during wet weather. These trails meet again after another half mile, so you can continue your hike either way without backtracking. The Underwood Trail has the added attraction of passing below some truly spectacular faces of metaconglomerate rock. This rock formed from the mixed sediments of ancient riverbeds and was subsequently thrust upward while being subjected to heat and pressure. Look carefully to see the embedded river stones and how they were stretched into ovals through this process.

Turn your attention now to the extremes on the Grandfather Trail, if you have chosen this route. You will ascend this section using ladders and cables that are bolted onto sheer rock. Pay attention and take care! The treeless communities you're scrambling across and over are high-elevation rocky summits. These differ from granitic domes in that the rock surface is fragmented rather than smooth, with steep rather than gradual slopes. Both attributes have implications for the plants that live there, though some plants grow in both natural communities. High-elevation rocky summits have fractured rock; the plants find soil in these small openings, but much of the surface is bare. Look at spots off the main path to get a better sense of the natural community.

Soon you'll come to a steep ladder with a sign that declares "McRae Peak, 5,939 ft." On a nice day, it makes for a great lunch spot; on rainy days, give the slick ladder and slippery peak a pass. There is very little vegetation here, but the views on a clear day are unforgettable!

You'll soon pass the second junction with the Underwood Trail. Not far past this point, at the 1-mi mark, you'll ascend Attic Window Peak, above an environment almost like an alpine meadow. This is another example of a place where wind is funneled through, keeping the area free of trees and allowing a profusion of alpine wildflowers to thrive in open sunlight.

Given the harsh conditions of these high-elevation communities at Grandfather Mountain, why do plants and animals live here at all? Surely the short growing season, frigid winter temperatures, wind, and ice are prohibitive to most living things. This assumption is supported by hard evidence. If you made a species list for these high peaks, it would be much shorter than a list for a similarly sized patch of northern hardwood forest or rich cove forest.

Yet there are still plants and animals that thrive in harsh environ-

Species such as sand myrtle form low-growing mats of vegetation ideally suited for life on the edge. Rocky summits provide niches for a variety of rare plant species and have more topographic variation than granitic domes.

ments. One explanation is that a species that can adapt to difficult conditions faces less competition. For example, the federally endangered Carolina northern flying squirrel (*Glaucomys sabrinus coloratus*) makes its home on Grandfather Mountain and just eight other high-elevation peaks above 4,500 ft. Why not lower elevations? It cannot compete for habitat and other resources with the common southern flying squirrel (*Glaucomys volans*) at lower elevations. The Carolina northern flying squirrel has found its unique niche at the transition between northern hardwood and spruce-fir forests, subsisting primarily on lichens and mosses that grow there.

Just before the 2-mi mark, the Grandfather Trail will end a little past Calloway Gap. If you have extra time or are hiking the Grandfather Trail as an out-and-back effort, continue another 0.3 mi to Calloway Peak, the highest peak in the Grandfather Mountain complex at 5,964 ft.

If you are completing the one-way trek down the Profile Trail, turn left and descend to Shanty Spring, following orange trail blazes. As the trail ducks below the ridgeline, it passes through spruce-fir forest that gradually transitions to a northern hardwood forest. Watch carefully for a transition from the sparsely vegetated rocky summit to an evergreen spruce-fir forest to the deciduous trees of the northern hardwood forest. The first third of a mile of this descent is rocky, steep, and eroded, a remnant of the old Shanty Trail, which has historic importance as the source of water nearest to the summit ridge.

Take your time as you descend and look for new species of trees and plants as you move into more protected environments. You'll soon notice the shiny, peeling bark of yellow birch (*Betula alleghaniensis*) and its long roots that allow it to perch, spiderlike, on top of the huge rocks on this section of trail, which passes through a boulderfield, a special subtype of northern hardwood forest. As you pass the seepy area of Shanty Spring, look for a large percentage of yellow birch in the canopy. Though other typical northern hardwood species grow here, such as yellow buckeye (*Aesculus flava*), yellow birch clearly dominates in these rocky environments.

Below Shanty Spring, you'll continue to descend until you reach the Profile View at 2.8 mi. This section of the Profile Trail, from here to the trailhead, was constructed with substantial thought and planning to protect the rare species and prevent erosion. Notice that these lower elevations and gentler slopes gain additional herbaceous species, and that these species differ depending on whether you are in a dry area, a seep, or somewhere in between.

The famous profile view of Grandfather Mountain is marked with a sign, so you can't miss it. The stone face was described in imaginative detail in *The Balsam Groves of Grandfather Mountain* (1907), by Shepherd Dugger, who used the flowery imagery of the early 20th century. You may smile at his overly romantic description, but the mountain that lies before you distinctively displays the furrowed brow and craggy features of an old, bearded man, the grandfather of Grandfather Mountain.

The next section of trail, below the Profile View and extending a third of a mile to a backcountry campsite at 3.1 mi, was constructed by hand in the late 1980s by Kinny Baughman and Jim Morton. Existing stones were moved manually into place, creating a smooth stone pathway and a massive stairway. This section is called Peregrine's Flight, and it is an exquisite feat of trail engineering. Be sure to visit

Extreme Botany

The term "extremophile" means "lover of extremes," in this case a plant or animal that "loves" extreme conditions. On this hike, you may have felt a bit like an extremophile yourself! You also saw several plant species existing in extreme conditions, like the flag-form spruce and fir hanging on for dear life in little pockets of soil on windy ridges, or the low-growing herbaceous species that thrive on rocky summits or in alpine meadows where conditions are too harsh for woody plants. Even rosebay rhododendron (*Rhododendron maximum*) is something of an extremophile, growing typically in acidic soils that provide insufficient calcium and other plant nutrients for many species.

From our field observations, we might conclude that extremophiles grow where they do because they "love" extreme conditions and will fail to grow under "better" conditions. However, this conclusion is inaccurate in at least some cases. Indeed, it can be shown that that some extremophiles (also called stress-tolerators) will grow quite nicely under less extreme conditions when planted and tended by humans. Just visit any arboretum, botanical garden, or even your own backyard to confirm this observation.

So why do we seem to find extremophiles only in extreme conditions in nature? In many cases, it appears that the extremophiles live under such difficult conditions because they cannot compete well for resources with other species that grow well *only* in favorable conditions. Our discussion of the northern flying squirrel (*Glaucomys sabrinus coloratus*) in this hike is a case in point. Another example of an extremophile is Fraser fir (*Abies fraseri*), which occurs naturally in the southern Appalachians only above

the campsite just off the trail, which also has some impressive stonework.

The undeveloped portion of Grandfather Mountain was purchased as a North Carolina state park in 2009 from the Morton family, using Natural Heritage Trust Fund monies and private donations. The attraction area is now run by a nonprofit organization, while the state park system monitors the undeveloped portions. The Profile Trail was almost never built because of plans for a 900-acre development. Grandfather Mountain has been designated a World Biosphere Re-about 5,000 ft. However, the species performs well at lower elevations in plantations (it is readily grown to elevations as low as 3,000 ft, and even lower if careful attention is paid to site conditions), to the delight of the growers who have made North Carolina the second-largest Christmas tree producer in the United States.

Thus some extremophiles may secretly be generalists, capable of growing in a rather wide range of conditions, including extreme conditions. If this is the case, these extremophiles are poor competitors and are excluded from the more favorable sites by better competitors; they are relegated to the extreme sites, where they live by default. It may be that extremophiles are really jacks-of-all-trades. However, like the proverbial jacks-of-all-trades, they are masters of none. When confronted by the fast-growing competitors that specialize on favorable sites, they simply fail to thrive. Ironically, then, it appears that the good competitors may indeed be the true specialists, "masters" of favorable sites but simply unable to grow on extreme sites.

An important ecological take-home lesson is that organisms are generally prevented from occurring on a particular site by one or both of two factors (setting aside, for the moment, dispersal limitation), the physical environment and unfavorable interactions with other species. When we observe that a particular species fails to thrive in a particular place to which it has access, we may jump to the conclusion that the physical environment is to blame. However, we cannot rule out the possibility that unfavorable interactions with other species may, in fact, be the real explanation for the poor performance of a species.

serve in part because of the numbers of rare species that make their homes on the craggy slopes.

Below the campsite the slope of the trail becomes gentler, especially below Foscoe View at 3.8 mi, and you will enter a lovely rich cove as the trail turns left. Notice how the narrow ravine opens up and the terrain forms a concave, bowl-like shape. Protection afforded plants by this landform and a deep and fertile soil are two important environmental factors that lead to the development of a rich cove forest, with a diverse wildflower understory.

As you enter the cove, you'll begin seeing typical rich cove species, such as some large basswood (*Tilia americana*) and black cherry (*Prunus serotina*), in addition to buckeye. As the understory becomes more diverse, it is notably less dominated by any particular species. The increase in species diversity hints at a subtle shift in the underlying soil pH, with several added species, such as black cohosh (*Actaea racemosa*), favoring the less acidic soils found here. In such environments, species must be adept at capturing site resources because they have to compete with many others that love these moist, nutrient-rich environments. An important attribute for these species, then, is their competitive ability to sequester space, moisture, and nutrients. If the variations in herbaceous diversity along this hike intrigue you, please read more about this topic in the "Wildflower Diversity in Temperate Forests" sidebar in hike 6 (Station Cove Falls).

At mi 4 you'll cross Shanty Spring for the last time. The trail crosses a small tributary of the Watauga River before it joins the larger river. Here you'll notice that at the lower elevations, the trail is dominated by the white-flowering rosebay rhododendron (*Rhododendron maximum*). This species, a member of the blueberry family, favors acidic soils, so its presence in these locations indicates a decrease in soil pH. Although you are still in a bowl-shaped cove, you now find yourself in an acidic cove forest dominated by eastern hemlock (*Tsuga canadensis*) and rosebay rhododendron. The dominating presence of rosebay rhododendron in the understory here leaves fewer resources and less space for herbaceous species; this structural difference is one of the primary distinctions between rich and acidic coves. However, the underlying reason behind this difference is soil nutrients and pH, directly resulting from properties of the underlying rock.

The trail eventually levels out and follows the Watauga River, leaving the forest at a kiosk that marks the trailhead and the parking area. It's hard to believe that such a strenuous hike was only 5 mi in length! If you have time, explore the attractions area of Grandfather Mountain when you drive back to get your other vehicle.

26. ELK KNOB

Why Are We Here?

Our ecological goal is to explore some of the variation in northern hardwood forest, a variable and fascinating high-elevation community of the southern Appalachians. At Elk Knob, you'll see three subtypes of northern hardwood forest. One of these is the typic subtype, featuring the "three Bs": American beech (*Fagus grandifolia*), yellow birch (*Betula alleghaniensis*), and yellow buckeye (*Aesculus flava*), plus sugar maple (*Acer saccharum*). You'll also see examples of two other subtypes: a boulderfield forest and a beech gap. Seeing them together will help you appreciate what makes them different.

You'll also be treated to a natural rock garden on your climb, along with an elfin beech forest with a lush grassy understory, which gives it a magical feel. The forest has a different flavor than others you have may have seen in the mountains, with fewer species of trees and a grassy, open understory dotted with wildflowers. Near the base of Elk Knob, where you start your hike, the trees are large and the canopy is rather open, with a very rocky forest floor, but one that is rich with early spring wildflowers. As you approach the summit, the canopy trees are reduced to large sapling size and are nearly 100% American beech. By the time you reach the summit, the beeches have shrunken into a gnarled form, growing as short, dense shrubs.

At the summit, you'll have excellent views to the south and north. To the south, you can see Long Hope Valley stretching out before you. To the north, you can glimpse Mount Jefferson, location of another of our hikes. Signage for both views from the summit will teach you the names of nearby peaks.

The Hike

Elk Knob, one of the newer parks in the North Carolina State Parks system, was initially purchased in 2002 by the North Carolina chapter of the Nature Conservancy in response to concerns among local citizens about possible development of the mountain. It was transferred

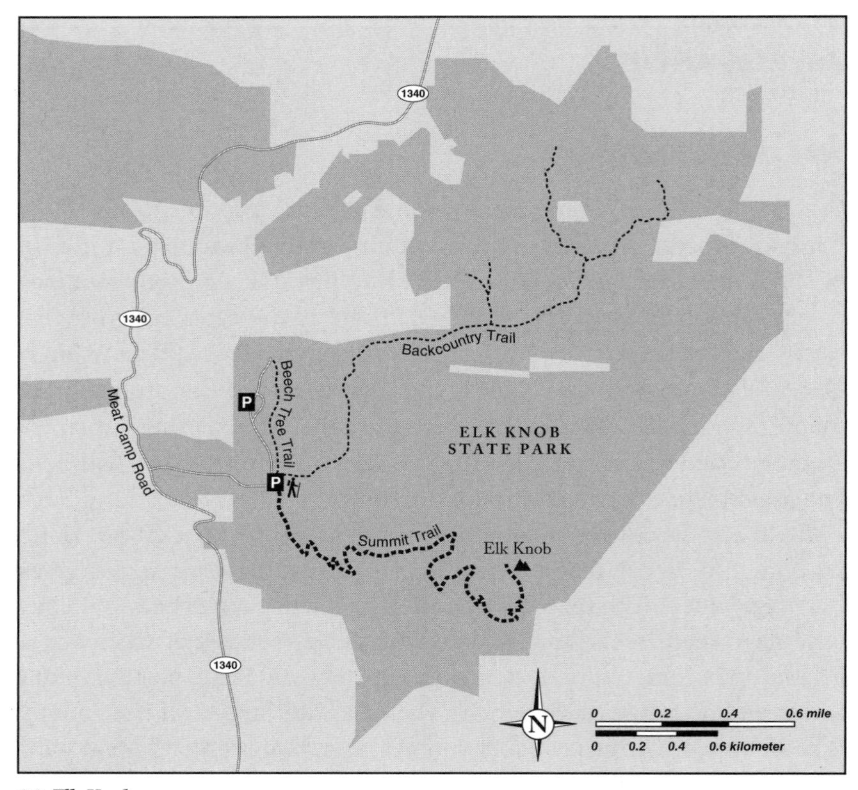

26. Elk Knob

to the state of North Carolina shortly thereafter, and it will remain undeveloped to conserve its scenic beauty and to help protect the watershed of the North Fork of the New River.

From the parking area, stop to register at the kiosk at the start of the hike, then follow the blue-blazed trail into the forest. You'll notice many exposed rocks and boulders on the forest floor along the trail. If your visit is timed for early May, you'll find that spring is just starting here. Elk Knob is a bit farther north than most of the other hikes in this book, and that, combined with its elevation above 5,000 ft, means that spring is both later and shortened, with early and later spring ephemerals blooming together. Thus you might see the nodding yellow trumpets of trout lily (*Erythronium* sp.), one of the earliest spring wildflowers, intermixed with flowers that typically bloom later, such as mayapple (*Podophyllum peltatum*), with its single white flower hiding

26. ELK KNOB

Highlights: A short hike through boulder gardens with diverse wildflowers, then to a dwarf beech forest with beautiful views from the top of Long Hope Valley

Natural Communities: Northern hardwood forest

Elevation: 4,600 ft (trailhead); 5,520 ft (summit)

Distance: 3.8 mi round-trip

Difficulty: Moderate, steep in places on this up-and-back climb to the summit

Directions: From US 221/421 in Boone, take SR 194N. After about 4 mi, turn left on Meat Camp Rd. and continue about 5.5 mi before turning right at the sign for Elk Knob State Park. The parking area for the trailhead is at the end of the entry road, about a half mile past the gate (accessible restrooms and picnic area are a left turn just before the trailhead).

GPS: Lat. 36° 19′ 37.9″ N; Long. 81° 40′ 36.8″ W (trailhead)

Information: Elk Knob State Park, 5564 Meat Camp Rd., Todd, NC 28684, (828) 297-7261

below paired umbrellalike leaves. Time your visit for early July and you'll be treated to the floriferous heads of Turk's cap lily (*Lilium superbum*) in patches of sunlight in the forest and in the sunny corridor of the old summit road.

At the start of your hike, the forest canopy is mixed. Some of the dominant canopy trees are sugar maple, northern red oak (*Quercus rubra*), American beech, black cherry (*Prunus serotina*), and white ash (*Fraxinus americana*). The largest trees are northern red oak, but as you climb up the western flank of Elk Knob, you'll notice a dramatic change in canopy composition. The oaks, along with black cherry, are common along the lower part of the hike but are absent once you ascend the slope, leaving a forest that is primarily American beech with pockets of yellow birch in the rockiest areas.

The natural community near the start of the hike could best be described as a rich cove forest, which has many variations. The lower slopes have a mix of deciduous canopy species, without a clear dominant, and the presence of white ash, black cherry, and basswood (*Tilia americana*) indicates a soil with a higher pH. But what's missing? Not a heath is to be seen, which tells you that these are not the typical acidic soils so common in the Blue Ridge. Even at the trailhead, the elevation is a bit high for tulip-tree (*Liriodendron tulipifera*), a species you'd expect to see dominating a rich cove. Here it's present, but in low numbers. The forest found at Elk Knob appears to be in pristine condition. Finding no cut stumps on the slopes, we can speculate that this forest is quite old—or at least has experienced limited disturbance—allowing the canopy and understory to develop over time.

Just past the half-mile mark, you'll cross the old road leading to the summit for the first of four times. Some species, such as Turk's cap lily, take advantage of the additional sunlight next to the road's edge, so look for additional wildflower species having a similar strategy if you visit between April and October.

As you climb through the switchbacks on the mountain, you may notice that the steepest and rockiest places are dominated by yellow birch, which thrives in these types of environments. This is a special type of northern hardwood forest called a boulderfield. Yellow birch is more adaptable to the challenges of shifting substrate and pockets of soil than other species, spreading its roots over and then into the soil found in pockets in the crevices. It's not the only species that can do this, but here it is the dominant. Boulderfields also tend to have abundant wildflowers, which thrive in microhabitats too small for trees.

Elk Knob is one of about a dozen mountain peaks in Ashe and Watauga Counties formed of rocks that make up the Amphibolite Mountains. These are part of the Blue Ridge Mountains, but the rocks are of volcanic origin. These igneous rocks, rich in minerals, were metamorphosed when the continental plates of Laurentia and Gondwana collided around 300 million years ago. The force of that collision, and several successive ones, pushed up the mountains known today as the southern Appalachians.

This same force transformed the existing volcanic rocks even as they were pushed up into mountains. Heat and pressure from the collision altered them into a metamorphic form called amphibolite. These rocks are distinguished from others by being unusually dark, flecked with black grains of the mineral amphibole, and they erode into less-acidic, rich soils with high levels of nutrients available for plant growth. These "sweet" soils provide habitat for several rare plant species at Elk Knob, such as Gray's lily (*Lilium grayii*).

You'll take a few switchbacks as you climb the western flank of Elk Knob. About a mile and a half from the trailhead, the trail will turn to the left and then begin a series of switchbacks up the north-facing slope. The rockiest, steepest sections, dominated by yellow birch, are tucked into larger areas that are nearly pure stands of American beech. This is another specialized type of northern hardwood forest called a beech gap. Here at Elk Knob, the north-facing slopes offer the cool and moist microclimate favored by American beech.

One of the first things you'll notice in the beech gap is that the trees are not large in diameter and grow very densely. Below them is a lush understory of sedges and grasses, most notably the narrow-leaved Pennsylvania sedge (*Carex pensylvanica*), with its tidy, combed look. In early spring, you will find the grassy understory dotted with large white trillium (*Trillium grandiflorum*), although compared to the lower slopes and rocky areas, the beech gap herb layer has fewer species. This could be explained by the sheer density of trees and the deep shade they cast during the growing season.

After more climbing and a few more switchbacks, you'll find yourself on the final ascent to the summit. You'll cross the old summit road, which has been closed to allow the forest to recover from severe erosion. This trail opened in 2011 after five years of construction, with thousands of hours contributed by volunteers and park staff. Take a rest on one of the stone benches and admire the carefully executed rock work along the trail.

Look around as you begin walking again, close to the summit. The forest is still primarily beech, but its character has changed again. Rather than growing straight and tall, the trees here are reduced to gnarled shrubs. When you reach the summit, you'll see that the beech forest is essentially a dense, shrubby hedge. This elfin form is sometimes called *krummholtz*, after German words that mean "crooked" and "wood." Other shrubs you'll see here include a few of what may be the first heath family species seen along this trail, minniebush (*Menziesia pilosa*), highbush blueberry (*Vaccinium corymbosum*), and flame azalea (*Rhododendron calendulaceum*). In addition, there is the beaked hazelnut (*Corylus cornuta*), blackberry (*Rubus* sp.), and skunk currant (*Ribes glandulosum*). There is no herb layer here at all; nothing can grow beneath the dense shrub layer. If you have visited a heath bald in the southern Appalachians, the summit here at Elk Knob looks similar, but decidedly less heathy.

*Beech grows in pure stands with a grassy understory on the northern slopes of Elk Knob. The trees grow straight and close together on the lower slopes (*bottom*), while the winds and harsh conditions of the upper slopes and summit reduce beech to a gnarled form called* krummholtz *(*top*).*

Northern Hardwoods—Natural Community or Crossroads?

Among the natural communities that we characterize in this book, none is more fascinating than the northern hardwood forest. Even its name is intriguing. Although three species that co-occur commonly to the north (American beech [*Fagus grandifolia*], yellow birch [*Betula alleghaniensis*], and sugar maple [*Acer saccharum*]) are often abundant in the northern hardwood forest, this natural community is hardly northern. Its climate is decidedly southern, with longer, warmer summers than one would encounter in hardwood forests to the north. Several of its diagnostic species (such as yellow buckeye [*Aesculus flava*]) are found only in the South.

Adding to further complexity is the variety of subtypes of northern hardwood forest. We recognize four subtypes in this book (typic, rich, beech gap, and boulderfield), but some authors prefer to recognize the boulderfield subtype as a distinctive natural community. Perhaps more than any of the other natural communities of the southern Appalachians, northern hardwood forest intergrades with a variety of other natural communities, including montane oak forest, spruce-fir forest, cove forest, grass bald, shrub bald, and rocky summit. The transition to cove forest is particularly gradual. This "crossroads" position has led some ecologists to view northern hardwood forest as nothing more than an extended transition zone among a variety of other high-elevation natural communities, not worthy of recognition as a natural community in its own right. However, we, like several other contemporary authors, prefer to consider northern hardwood forest a distinctive natural community.

Variation within the concept of northern hardwood forest is substantial. Like cove forest, northern hardwood forest is responsive to variation in soil fertility. Stands on acidic parent materials represent the typic subtype, with a relatively species-poor sedge- and grass-dominated understory. On mafic rocks, like those at Elk Knob, the high-fertility soils support the rich subtype of northern hardwood forest with its lush, species-rich understory.

Two subtypes of northern hardwood forest are especially perplexing, challenging ecologists to unravel the forces shaping their composition and distribution. Embedded within spruce-fir forest, the beech gap subtype represents a high-elevation anomaly. Why doesn't spruce-fir forest simply extend across the high-elevation gaps occupied by these nearly pure stands of American beech? Perhaps winter winds carrying their burdens of snow and ice up the valleys and across low points along ridges select for a hardy deciduous species that is entirely leafless during the harsh winter.

The boulderfield subtype of northern hardwood forest is also quite unusual. Occupying tumbled masses of boulders that fill high-elevation ravines, the boulderfield subtype dominates this unusual habitat because of yellow birch's unique ability to germinate and establish itself successfully on moss-covered rocks. It later sends its roots snaking amongst the boulders in search of the more reliable sources of water and nutrients afforded by mineral soil. There is even uncertainty as to the origin of the boulderfields themselves. It seems reasonable to assume that boulders were tumbled downslope and collected in high-elevation ravines during a colder period, when frequent freeze/thaw cycles caused the formation of ice wedges that pried boulders loose from the soil, sending them careening into the ravines. We have little direct evidence of the precise forces that created these spectacular boulderfields.

Finally, it is interesting to note that paleoecologists have found little evidence of a northern hardwood forest during the last glacial maximum. While other natural communities, such as spruce-fir forest and montane oak forest, were apparently common during the last Ice Age, northern hardwood forests are a uniquely postglacial phenomenon. They seem to have assembled fairly recently from elements of the flora that existed in refugia in the Coastal Plain and Piedmont, combined with other hardy elements that remained in place at high elevations during the Ice Age.

At the summit of Elk Knob, you'll have the option of turning left to take in views to the south, which include Grandfather Mountain, or right to see summit views to the north, including Mount Jefferson. Choose one, but be sure to spend time at both summit viewpoints, a rich reward for your hard work in reaching the top. On your way down, note the changing structure of the canopy as you walk through three kinds of northern hardwood forest.

27. MOUNT JEFFERSON

Why Are We Here?

Until recently, Mount Jefferson was one of the few public lands in the northwestern part of North Carolina. It was designated a state park in 1956 after years of effort by local residents to secure the funds and acreage needed. This isolated mountain rises 1,600 ft above the surrounding valleys and communities of Jefferson and West Jefferson.

The ancient oak forest here tells an interesting story of the resilience of our southern Appalachian forests and their vulnerability to outside influences. You'll hear whispers throughout your journey that tell of something missing. American chestnut (*Castanea dentata*), formerly considered the redwood of the East, has left lasting reminders of its former prominence. As you hike Mount Jefferson's trails, you'll look for clues to this story.

Although other forest types are present here, the montane oak forest is most notable as a fine example of its type. Mount Jefferson was designated a national natural landmark in 1974 partly because of this unharvested forest. Though gnarled by the harsh weather at more than 4,000 ft of elevation, it stands as a quiet witness to a story of forest change.

The Hike

From the parking area, walk up a paved trail through the picnic area. Restrooms are down a short spur to your right. You'll see the Rhododendron Trail spur to your right, which you'll hike on your return trip. For now, continue straight and follow the road. It's just a short, steep, third-of-a-mile climb on this gravel road to the Mount Jefferson summit, with views along the way, and there is a spur trail (marked with a diamond-shaped white hiker blaze) to your left that leads a short distance to the summit area. Though the views on a clear day are nice, the summit is less than spectacular, marred by a big radio tower. You'll want to hike to the summit because it's there, but a quick look through the forest windows to the surrounding countryside is all

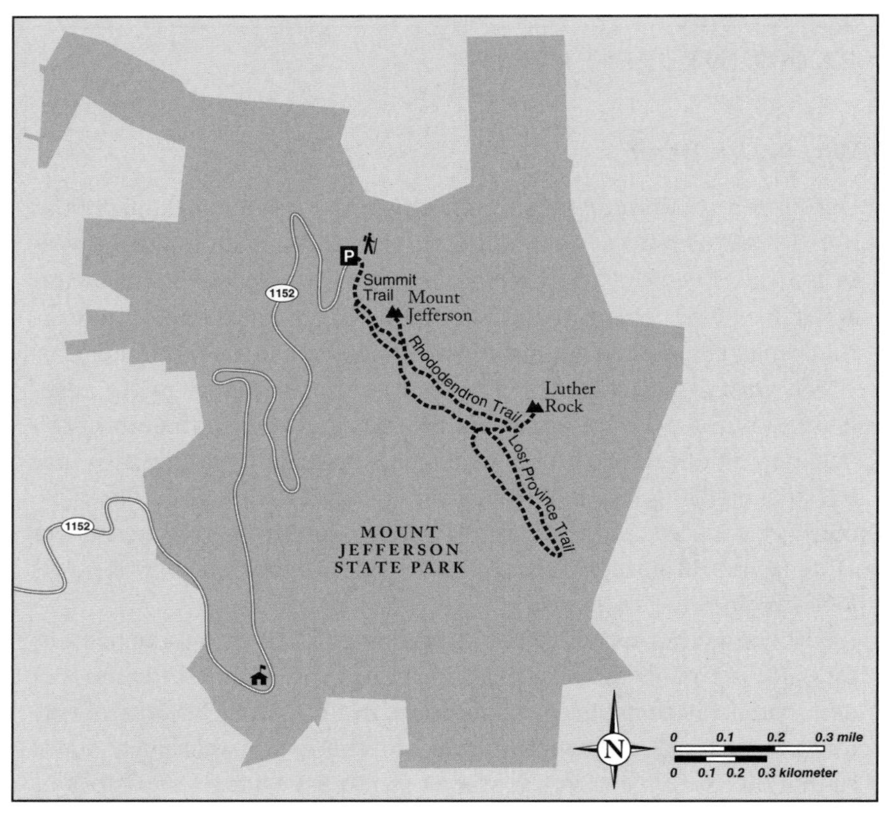

27. Mount Jefferson

you'll need. Descend the summit trail and turn left onto the Rhododendron Trail, which is marked with red circular blazes.

You'll have views along the trail to the north side of Mount Jefferson on your way to Luther Rock, about another third of a mile. You're walking along the ridgeline of Mount Jefferson, and the forest here is exposed to the elements. Although there are some stands of northern hardwood forest on the north-facing slopes, which stay cooler and moister during the hot Carolina summers, most of the forest is montane oak forest. The predominant species are northern red (*Quercus rubra*) and white (*Q. alba*) oaks, with red oak clearly the more dominant. The predominance of northern red oak might be caused by the nutrient-rich soils derived from the amphibolite rock that forms the mountain.

27. MOUNT JEFFERSON

Highlights: An ancient oak forest and views of the surrounding countryside

Natural Communities: Montane oak forest, northern hardwood forest, shrub bald

Elevation: 4,665 ft (Mount Jefferson summit)

Distance: 2.2 mi round-trip (two connected loops)

Difficulty: Easy to moderate

Directions: From the US 221 Bypass between Jefferson and West Jefferson, look for signs for Mount Jefferson State Park. Coming from the south, turn right on NC 1152 (from the north, turn left) and follow signs to the parking area (accessible picnic area, though restrooms are not accessible; plans are under way for a new, accessible restroom facility).

GPS: Lat. 36° 24′ 11″ N; Long. 81° 27′ 48″ W

Information: Mount Jefferson State Park, 1481 Mount Jefferson State Park Rd., West Jefferson, NC 28694, (336) 246-9653

The ridgeline of Mount Jefferson runs in a northwest to southeast direction, and it separates two major watersheds. To the north (the left side of the trail), water flows to the north branch of the ancient New River. To the south (the right side of the trail), water flows to its south branch. Mount Jefferson is part of the ancient chain of the Amphibolite Mountains that includes Elk Knob (hike 26).

If you visit in late May or early June, you'll see the clusters of pale pink flowers of mountain laurel (*Kalmia latifolia*) and the magenta blossoms of Catawba rhododendron (*Rhododendron catawbiense*) on the drier, open sections of the trail, to your left. This is the high-elevation species of rhododendron, with rounded leaf bases. Along the ridgeline, where the forest is exposed to harsh winters, the trees are rather gnarled and bent. In some places, the Catawba rhododendron reaches into the canopy.

When you reach the Luther Rock spur trail, turn left and walk a few hundred feet to the amphibolite rock outcrop. Enjoy the views, which are much better than they are at the summit. You can look back to your left to see the ridgeline you just walked and the radio tower that marks the summit. Notice the open but continuous forest along these northeast-facing slopes. If you are visiting in the summer months, look carefully below the summit to see if you can spot canopies with round leaves that shake in the slightest of breezes. Mount

Even before the trees leaf out, the generous spacing among chestnut oaks at Mount Jefferson indicates the long-term consequences of losing a former canopy dominant, American chestnut.

Jefferson has a rare population of big-toothed aspen (*Populus grandidentata*); this species is much more common to the north.

If you walk around to the right side of the Luther Rock outcrop, you'll note that the vegetation here is more open, with much exposed bare rock interspersed with mountain laurel and Catawba rhododendron, as well as highbush blueberry (*Vaccinium corymbosum*). The low-growing shrubs are able to withstand the drought associated with the southwest orientation and shallow soils. Although there seem to be enough oak and other trees on this side of Luther Rock to qualify this as a montane oak forest, some would argue that the natural community here is more like a shrub bald.

As you view the slopes below, note the openness of the canopy. Far from forming a tight, closed-in forest, the gnarled oaks that make up most of the forest canopy have gaps between them. These trees, although short and bent because of the harsh conditions on the ridgeline, nevertheless comprise a fine old-growth example of montane oak forest. Logging and development have never marred the summit of Mount Jefferson.

Where Has the Chestnut Gone?

American chestnut (*Castanea dentata*) once made up as much as 25% of the canopy in forests that are now dominated by oaks. Valuable to both humans and wildlife, chestnut played an important role in the forests of eastern North America through the middle of the 20th century. Chestnut's strong and easily workable wood survives to this day in furniture and historic structures. The reliable and abundant crop of nutrient-dense chestnuts provided food for black bear (*Ursus americanus*), white-tailed deer (*Odocoileus virginianus*), wild turkey (*Meleagris gallopavo*), and the now-extinct passenger pigeon (*Ectopistes migratorius*). Unlike oaks, which have fluctuating cycles of acorn production, calorie-dense chestnuts were produced abundantly, with estimates of 6,000 chestnuts produced per year per mature tree. Although the historic range of American chestnut reached north to Maine and west to the Mississippi River, it had its greatest presence in the southern Appalachians.

American chestnut disappeared with stunning rapidity after the introduction of the chestnut blight, a fungus (*Cryphonectria parasitica*) that was first identified in New York City on a cultivated Asian chestnut in 1904. A relatively harmless parasite of Asian chestnut, the chestnut blight proved lethal to the American chestnut, which had not coevolved with this organism. As American chestnut trees matured and fissures opened in the bark, blight spores alighting on the thin bark caused cankers to form, which eventually girdled the trees. Within 3 years of initial infection, the trees died.

By 1950, more than 4 billion chestnut trees stood dead, their rot-resistant trunks continuing to stand as silent witnesses for decades. The trees weren't entirely dead, however; they continued to sprout from the stumps and roots, because the carbohydrate stored in the deep root systems generated fresh shoots. These shoots are usually killed by the blight before they reach sexual maturity, and because they are clones of the original tree, they have no resistance to the blight. One might wonder how the chestnut blight persists in these forests in the absence of large, mature chestnut trees. In a quirky twist of fate, the fungus causing chestnut blight can persist on a variety of oak species and is thus available at almost any time to infect emerging chestnut sprouts. Occasionally, however, chestnut sprouts will reach maturity and produce their spiky-husked fruits, offering a sliver of hope that resistance may be found in the genetic recombination associated with sexual production of seed.

Those at the American Chestnut Foundation have a vision of restoration of the American chestnut tree in eastern forests; through decades of work and research, they have created a hybrid that is 15/16ths American chestnut with the blight resistance of Chinese chestnut. These new hybrids are undergoing test plantings in the hopes that they can be used to restore this giant of eastern North American forests.

Why do we mourn the loss of the American chestnut? The forest in which you stand is well over a hundred years old, and it seems healthy enough. The canopy is more open, and chestnut sprouts live on, albeit briefly. However, wildlife biologists believe that many animal populations have been reduced by the loss of this bountiful, once-ubiquitous tree. Ultimately, the demise of American chestnut reminds us of the broad and tragic consequences of introducing exotic pests.

After you have explored Luther Rock, retrace your steps and rejoin the Rhododendron Trail to the left, which goes only a short way before it intersects the Lost Province Trail. At this junction, turn right to descend the southeast-facing ridgeline. The trail then loops around and heads northwest along the flank of Mount Jefferson before rejoining the Rhodendron Trail to complete a three-quarter-mile loop.

One thing you should notice as you walk back along the lower, southwestern-facing slope is the composition of the forest canopy. You are in a thick montane oak forest, one that has a canopy shared by northern red and white oaks as well as chestnut oak (*Quercus montana*). Chestnut oak tolerates these drier conditions and hot summer sun on the southwest-facing slope. Look for its fallen leaves, with their scalloped leaf edges, and its sizable acorns along the trail at your feet.

Interpretive signs may call this forest an oak-chestnut forest, but they aren't referring to chestnut oak. Rather, they are referring to American chestnut. There is speculation that the forest canopy is open here because nothing has perfectly filled the niche of the chestnut. Look up through the canopy to determine how much sky you can see through the trees, and imagine these gaps filled by the canopies of the once-mighty chestnut.

As you continue your hike, look for remnant sprouts of American chestnut, which are common, never having been damaged by salvage logging. The woody shoots coming from old stumps will have long, elliptical, smooth leaves with sharp teeth along their margins. Similar but smaller hairy leaves indicate the American chestnut's little cousin, chinquapin (*Castanea pumila*).

When you reach the junction with the Rhododendron Trail, take the fork on the left to connect to and hike its lower section to complete the loop, then turn left to walk through the picnic area to the parking lot. All told, the two loop trails and spurs are about 2.2 mi long. As you hike out, imagine again what this forest might look like if the American chestnut were still here as a dominant canopy tree. The openings to the sky and diminished sprouts from old stumps are a poignant reminder of the unintended consequences and far-reaching impacts of some human activities.

28. WHITETOP MOUNTAIN

Why Are We Here?

Open meadows with a stunning mountain backdrop have always been popular destinations, and Whitetop Mountain, as Virginia's highest peak, is no exception. The Whitetop Mountain Folk Festival attracted thousands of people to its cool summits from 1932 to 1939; as many as 20,000 may have attended in 1933 when Eleanor Roosevelt included the folk festival on her mountains tour. Like them, you are here to enjoy the open meadows and views. You might even feel compelled to burst into a chorus from *The Sound of Music* when you're up there!

You'll climb from Elk Garden on the Appalachian Trail through sections of northern hardwood forest, emerging into a grassy meadow just before you reach the dirt road that winds toward the summit of Whitetop Mountain. Here you'll see the clear delineation of natural communities, from northern hardwood forests to pure stands of red spruce (*Picea rubens*) to open grass balds. In this mosaic of high-elevation natural communities, we'll explore questions about the origins

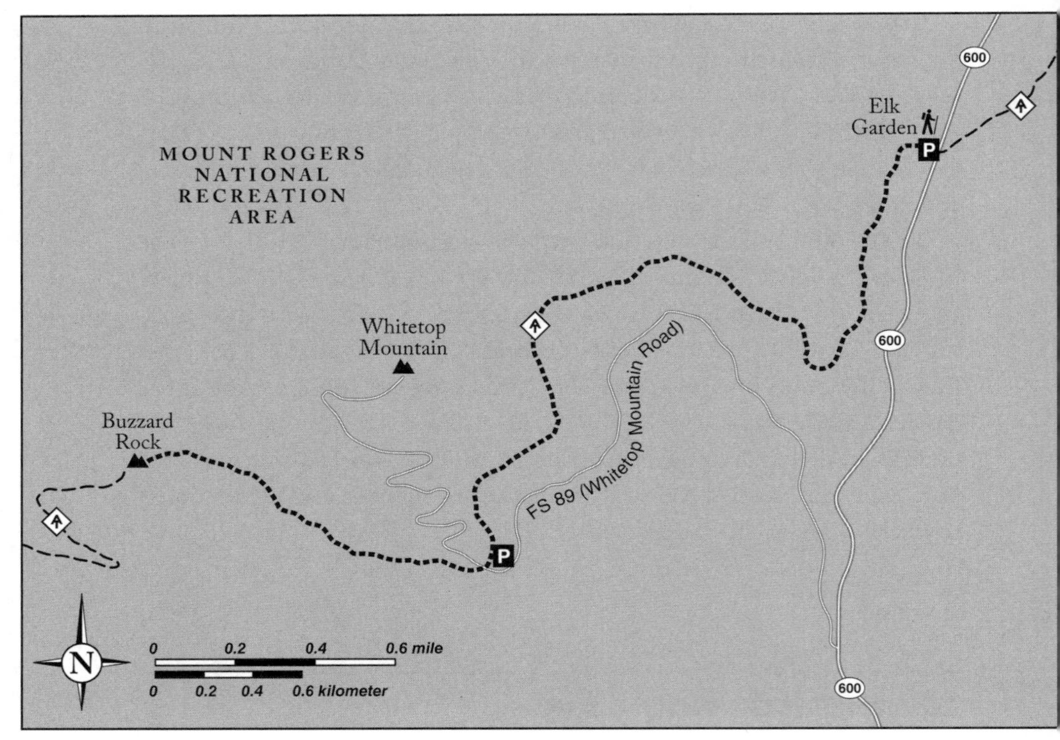

28. Whitetop Mountain

of grass balds, why the spruce-fir forest is really just a spruce forest, and what tiny alpine plants can tell us about the past history of this area. At Buzzard Rock, you'll be rewarded for your efforts by unparalleled views of the surrounding mountains on clear days.

The Hike

At the parking area at Elk Garden, review the area maps at the kiosk. Please take time to sign the logbook, even though you are day-hiking, noting that you are southbound to Buzzard Rock and back. With the road and kiosk behind you, you'll see the southbound Appalachian Trail, marked by its distinctive white rectangular blazes, disappearing into the woods. That is where you'll start your journey.

The first 2 mi are a steady climb through lovely northern hardwood forest. Since you are near the northern end of the southern Ap-

28. WHITETOP MOUNTAIN

Highlights: Open meadows and a grassy bald below the summit of Virginia's second-highest peak; lovely views from Buzzard Rock

Natural Communities: Grass bald, northern hardwood forest

Elevation: 4,434 ft (Elk Garden trailhead); 5,520 ft (Whitetop summit)

Distance: 6.5 mi round-trip; does not include a 2-mi round-trip spur option to the summit of Whitetop Mountain. For an easier 1.5-mi round-trip, park on the Whitetop summit road next to the Appalachian Trail and walk to Buzzard Rock and back.

Difficulty: Moderate

Directions: From Damascus, VA, drive east on US 58E past the community of Whitetop. Turn left on VA 600. Drive 2.9 mi; Elk Garden parking area will be on your left. For the shorter option, make the same left turn on VA 600 from US 58E. Drive 1.6 mi, then turn left on FS 89 (Whitetop Mountain Rd.) and drive until you see a small parking lot on your right. The Appalachian Trail crosses the road here; from the parking lot, cross the road and join the route described below.

GPS: Lat. 36° 38′ 47.2″ N; Long. 81° 34′ 59.4″ W (Elk Garden trailhead)

Information: Mount Rogers National Recreation Area, 3714 Highway 16, Marion, VA 24354, (276) 783-5196

palachians, the climate is a bit cooler. That translates into a shift in species composition to lower elevations. Here that means that more cold-tolerant species can be found in the northern hardwood forest.

As you climb, look for yellow birch (*Betula alleghaniensis*), the most common species, with its shiny, peeling bark and horizontal lenticels. Other common hardwood species include sugar maple (*Acer saccharum*) and American beech (*Fagus grandifolia*). Look for an evergreen tree here as well, red spruce. Spruce typically creeps into the higher-elevation examples of northern hardwood forests, so you might think that, based on its presence, you are well over 5,000 ft. In fact, you are not; Elk Garden is at 4,400 ft, and where you cross the road going to the summit of Whitetop, you'll have just reached 5,000 ft. Higher latitudes mean that the transition to spruce-fir forest occurs at lower elevations.

The consequences of a generally cooler climate at this latitude are subtle, but you can find other clues to support this idea. Another species you may see that helps tell the story is witch hobble (*Viburnum lantanoides*), a shrub with large, rounded, deeply veined leaves that are positioned in opposite pairs along their branches. In places like Mount Mitchell in North Carolina and Clingman's Dome in Tennessee, this species occurs primarily in spruce-fir forest above 5,000 ft. Here it grows at lower elevations in northern hardwood forests. Its presence here reinforces the idea that this species is tied to its own environmental needs, rather than a particular natural community.

You will pass a backcountry campsite on your left as you start seeing openings into the treeless meadows soon to come. Shortly thereafter, you'll emerge onto an open pathway with the forest to your left and open meadows stretching up to your right. Pause here and look up—you should be able to see the winding dirt road to the tree-covered summit of Whitetop Mountain.

When you reach the road at 2.3 mi that winds toward Whitetop Mountain, you can opt to make the mile-long trek to the summit of Virginia's second-highest peak. If you are going for the views, stop at the parking lot just below the summit; the actual summit is in the trees, which share space with some large radio towers. As is the case at Mount Rogers, you will find the actual summit shrouded in evergreen forest. Interestingly, the summit of Whitetop supports a pure stand of red spruce, containing no Fraser fir. In the sidebar, we'll explore some possible reasons for the absence of fir.

After you cross the road that leads to the Whitetop summit, start looking for shallow, rocky areas along the trail. Examine them closely and look for clumps of plants that are trifoliate—that is, the leaves have three small, toothed leaflets each. Some of these may be bright red, and you may see tiny white flowers on them in late June. Three-toothed cinquefoil (*Sibbaldia tridentata*) may not look especially impressive, as wildflowers go; however, it is the only member of its genus, and its nearest relatives find their homes in alpine meadow environments to the north—meadows that occur above treeline. Ecologists consider cinquefoil's presence concrete evidence of the natural origins of the grass bald natural community on the slopes of Whitetop out to Buzzard Rock.

Continue through the lovely open meadows, ducking into northern hardwood forest once again briefly before emerging into the open for the rest of your hike. From Whitetop Road to Buzzard Rock, it's about three-quarters of a mile. Ahead of you, look for a very dark rock outcrop—this is Buzzard Rock, your destination and turnaround

Three-toothed cinquefoil, a glacial relict more common to the north, grows abundantly on top of exposed rock outcrops on Whitetop Mountain. You can also find it on hikes 19 (Craggy Pinnacle) and 24 (Grassy Ridge).

point. Follow the white rectangular blazes on posts that mark the trail through the open meadows and up to Buzzard Rock. It's a perfect spot to enjoy the views and have lunch. If you'd like to know more about grass balds, refer to the "Grass Bald Origins and Future" sidebar in hike 13 (Hooper Bald). The next hike in this book, hike 29 (Mount Rogers), also has a sidebar, "When a Grass Bald Isn't," which explains that the "bald" area beneath Mount Rogers differs markedly from that on Whitetop.

From your buzzard's-eye view, you can clearly see the transitions between different forest types on the slopes below the Whitetop summit. The open grass balds are outlined by the deeper greens of the nearly pure evergreen spruce forest. Below, the lighter green northern hardwood forests have canopies with a softly rounded appearance because of the greater openness and diversity of tree species. This diversity becomes especially obvious in the fall when the deciduous trees change color.

When you've had your fill of the views from Buzzard Rock, prepare for the easy descent back to Elk Garden. On your return, pay attention to the transitions between open meadows and deep forests; unlike

What Dictates Plant Distribution?

There are two are useful premises to consider when assessing the differences between Whitetop Mountain and Mount Rogers, the two tallest peaks in Virginia. The first states that the presence or absence of any species can be tied directly to its environmental requirements. However, we have seen exceptions throughout this book, so another explanation is needed. The second is that history and geography also determine the occurrences of a species. Jeff Nekola (University of Wisconsin—Green Bay) and Peter White (University of North Carolina—Chapel Hill) presented these two ideas as conceptual pillars for explaining plant distribution. We will use them to explore the reasons for the presence and absence of two species found on either Whitetop Mountain or Mount Rogers, but not both. Fraser fir (*Abies fraseri*) is found on Mount Rogers but is notably absent on Whitetop Mountain, while a tiny grass bald indicator species (three-toothed cinquefoil [*Sibbaldia tridentata*]) is found on Whitetop but not on Mount Rogers.

The first explanation would assert that these mountains differ enough environmentally that Fraser fir is unable to grow at Whitetop. It seems unlikely that the Mount Rogers summit, just 200 ft higher and fewer than 20 mi away, offers a significantly different climate than Whitetop, so we can safely assume that variables like temperature and moisture are similar. Geological studies in the area have revealed that the soil-forming rocks on both mountains are volcanic in origin and similar. You could test the first explanation with a classic garden experiment, such as the one discussed in hike 24 (Grassy Ridge), and plant Fraser fir on Whitetop Mountain. Our guess is that fir would thrive, as would the cinquefoil if it were planted on Mount Rogers. So we must consider a second explanation for these paradoxes.

One plausible explanation for the absence of Fraser fir on Whitetop lies in the fluctuating climates of the planet's distant past. When the climate warms, species requiring cooler temperatures migrate to higher elevations. Trees rooted at lower elevations die off, while seedlings thrive in the cooler temperatures afforded by the upper slopes. Over time, the population moves upslope. However, trees such as Fraser fir are already clinging to the highest summits and have no place to go when temperatures increase. It's possible that Whitetop Mountain had a population of Fraser fir that was pushed right off the summit when the climate warmed sometime in the past.

However, Mount Rogers may have had a refuge—a steep, north-facing slope, for example—where a few trees or seedlings survived, or perhaps where a few seeds lay dormant in the seed bank. When the climate shifted again, bringing cooler temperatures back to these mountains, Mount Rogers had a source population of Fraser fir that could spread and recolonize. Whitetop, with no local population available, became reforested with red spruce (*Picea rubens*). You can learn more about colonization limitations on red spruce and Fraser fir distributions in the sidebar in hike 3: "The Case of the Missing Spruce."

A different explanation based on history and geography suggests a climate-related hypothesis for the presence of three-toothed cinquefoil on Whitetop. Ecologists believe that the presence of three-toothed cinquefoil in these meadows is an artifact of climate change. When cooler temperatures and harsher winters pushed trees like red spruce and Fraser fir off the summits, Whitetop and Mount Rogers may both have been alpine meadows well above treeline. However, this explanation still does not account for the absence of three-toothed cinquefoil on Mount Rogers. In the next hike, hike 29 (Mount Rogers), you will cross the road and begin your hike toward Virginia's highest peak, and you will not encounter three-toothed cinquefoil anywhere along your way, though you will pass open habitats that could certainly support it. Here, history again offers the most plausible explanation. Destructive logging and fires in the 1900s on the slopes below the peak eliminated the local population.

In the case of vegetation differences between Whitetop Mountain and Mount Rogers, using either environmental factors or history and biogeography to explain what we see is not sufficient. It is the combination of the two that provides the best explanation for what we see on our hikes.

most forest transitions, the boundaries between these communities are easy to see. Pretend you are a tree seedling as you walk through each community type, noting differences between the two environments. The cool, moist, shady forest would provide a home very different from the sunlight, wind, and desiccation of the open meadow. The stark contrast between environments of the forest interior and

From Buzzard Rock, look back the way you came and toward the summit of Whitetop Mountain. Natural community boundaries—between the open grass bald; dark, pointed red spruce canopies at the summit; and lighter green, rounded canopies of northern hardwood forest—are very evident from your perch.

meadow may keep the community boundaries sharp. We may not understand all of the reasons for the distinct boundaries between contrasting natural communities, but it's easy to understand why so many visitors are attracted to this beautiful landscape.

29. MOUNT ROGERS

Why Are We Here?

Many hikers deem this stretch of the Appalachian Trail to be among the most scenic. Here the Appalachian Trail climbs through beautiful northern hardwood forests of the Lewis Fork Wilderness Area for most of the first half of the hike, with occasional openings into meadow vistas improbably dotted with shaggy wild ponies. Just when the summer sunshine begins to feel hot, you'll take the Mount Rogers spur trail and enter the cool and shady spruce-fir forest on your way to the

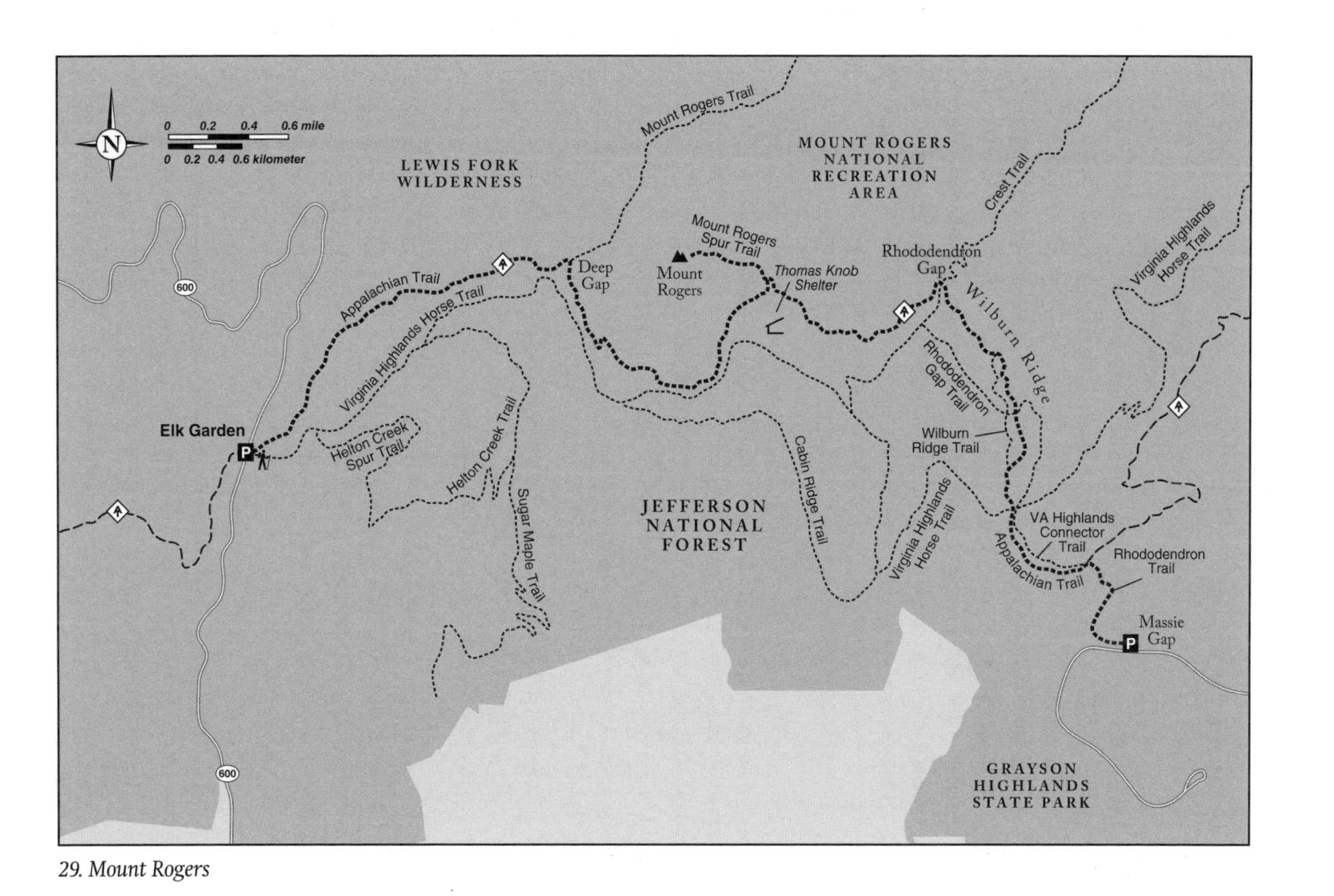

29. Mount Rogers

29. MOUNT ROGERS

Highlights: A breathtaking point-to-point hike along the Appalachian Trail to the summit of Virginia's highest peak, with opportunities to see wild ponies

Natural Communities: Northern hardwood forest, spruce-fir forest

Elevation: 4,434 ft (Elk Garden trailhead); 5,729 ft (Mount Rogers summit); 4,650 ft (Massie Gap)

Distance: 9 mi point-to-point, which includes a 1-mi round-trip spur trail to the summit of Mount Rogers

Difficulty: Moderate; steady climbing to Mount Rogers, then descending to Massie Gap

Directions: Since this is a one-way hike, you will need to leave a vehicle at Massie Gap in Grayson Highlands State Park. From Damascus, VA, drive US 58E (US 58W from Independence, VA) to the turnoff for Grayson Highlands State Park at VA 362. There is a small day-use fee to enter the state park. Drive up the park road, and the parking area for Massie Gap will be on your right. Park one car there, then drive the second vehicle back down to the park entrance and turn right (west) onto US 58. Travel 7.5 mi and turn right onto VA 600. Drive 2.9 mi; Elk Garden parking area will be on your left.

GPS: Lat. 36° 38′ 47.2″ N; Long. 81° 34′ 59.4″ W (Elk Garden trailhead)

Lat. 36° 38′ 0.1″ N; Long. 81° 30′ 32.5″ W (Massie Gap trailhead)

Information: Mount Rogers National Recreation Area, 3714 Highway 16, Marion, VA 24354, (276) 783-5196

Grayson Highlands State Park, 829 Grayson Highland Lane, Mouth of Wilson, VA 24363, (276) 579-7092

highest point in Virginia, Mount Rogers, at 5,729 ft. The descent from the summit into Grayson Highlands State Park offers additional opportunities to see wild ponies and consists mostly of open meadows with scattered rock outcrops and outstanding views. You will also pass by the Thomas Knob shelter, one of the nicer shelters on the Appalachian Trail, where you might have the opportunity to chat with backpackers hiking a longer section or through-hikers heading north to Maine. If you want to create some so-called trail magic, carry some extra snacks to share with any through-hikers you meet.

The Hike

Sign the register at the kiosk at Elk Garden (noting your direction and destination) before crossing VA 600 and walking through the turnstile gate to head north on the Appalachian Trail, following the white rectangular blazes. You will ascend through a pasturelike area before entering the forest. You may see the grazing cattle that keep the turf short here, or other evidence of bovine activity. Here you'll also see the blue-blazed Virginia Highlands Horse Trail heading off to the right; this trail will intersect the Appalachian Trail at several points along your hike.

At 0.6 mi you'll reach another gate, where you'll enter the Lewis Fork Wilderness. The forest here is typical northern hardwood forest, dominated by sugar maple (*Acer saccharum*), yellow birch (*Betula alleghaniensis*), and white ash (*Fraxinus americana*). Notice that none of the trees is particularly large. The railroad that runs from Whitetop Station to Damascus was used to clear all the mature timber from this region in the early 1900s, so this forest is not especially old. The 200,000-acre highlands area was designated as Mount Rogers National Recreational Area in 1966.

After 2 mi, you'll see signs that prohibit camping for the next 500 ft, and you'll arrive at another junction with the blue-blazed Virginia Highlands Horse Trail at Deep Gap. Continue straight, following the white rectangular blazes of the Appalachian Trail.

At 2.2 mi, take a sharp right turn and continue to follow the white rectangle blazes past a large boulder to your left. You'll start climbing a gentle grade toward Mount Rogers, and you'll notice that the forest is changing. Start looking for large evergreen trees scattered throughout the northern hardwood forest. These are red spruce (*Picea rubens*), which are more common in northern latitudes. As you ascend, you are entering a transition zone between northern hardwood forest and evergreen spruce-fir forest. Because each species in these communities has its own specific environmental requirements, these natural communities overlap and the transition is gradual.

About 3 mi from the trailhead and around 5,300 ft, you'll pick up a second needle-leaved evergreen tree, Fraser fir (*Abies fraseri*). As soon as you see the first one, examine the next 10 evergreen trees you see. How many are spruce and how many are fir? Keep this ratio in mind; you'll compare this ratio with one you'll find farther on.

Fir is friendly. Spruce is sharp. Try "shaking hands" with a branch, and you'll immediately appreciate this difference. Evergreen Fraser fir and red spruce look similar to the untrained eye, but it is easy to learn to recognize the differences between them. First, grab a branch and pull off a few needles. Fir needles are flat with blunt tips, while spruce needles are round and pointed. Try rolling needles between your fingers. Since fir is flat, it will not roll; spruce will. Examine the branches and trunk to see that fir bark is smooth with blisters of sap (which can be "milked," giving it the Appalachian name "she-balsam"), while spruce branches are bumpy. Finally, look for cones at the tops of the trees. Fir cones perch on top of the branches like candles on an old-fashioned Christmas tree, while spruce cones hang underneath the branches.

While Fraser fir and red spruce may appear similar and co-occur in many areas, their geographic ranges and ecological requirements differ slightly. Fraser fir is a southern Appalachian endemic, meaning it only occurs in the southern Appalachians. Fraser fir's close relative, balsam fir (*Abies balsamea*), is found from the northern Appalachians north to New England and from the upper Midwest and across most of Canada as far west as Alberta. Red spruce, on the other hand, is widespread, occurring in the southern Appalachians only at the very southern end of its natural range. It extends up the Appalachians to New England, the Canadian Maritimes, and Ontario. Ecologically, Fraser fir tolerates conditions found at higher elevations, while spruce outcompetes fir in the lower part of the spruce-fir elevation zone. In the middle, you'll find a mix of both species.

Shortly after encountering your first Fraser fir, you'll see forest openings to your right; you can cut through any of a number of short paths to peek into the open meadow with its gorgeous mountain backdrop, or you can continue another tenth of a mile until the trail comes to a fenced area. There are two herds of wild ponies that roam these southwestern Virginia highlands. If you are lucky, you may see one of the herds grazing in the rocky meadow. The ponies are on the small side and shaggy. Their grazing in the highland mists or on a clear day with a mountain backdrop makes for a picturesque setting. The horse trail will go right and through the turnstile into the meadow; foot travelers will continue to the left on the Appalachian Trail and head back into the forest.

Just under 4 mi from the trailhead, you will exit the forest once again, and soon thereafter you'll reach the blue-blazed spur trail that leads to the left to the Mount Rogers summit. If the weather is souring

The open, rocky landscape along the Appalachian Trail spur into Grayson Highlands State Park was created by extensive logging followed by damaging slash fires in the early 1900s. The grassy areas also have a clipped appearance, caused by grazing from the herds of wild ponies in the state park.

or if it's late in the day, you can bypass this 1-mi round-trip spur trail and continue on the Appalachian Trail toward Grayson Highlands State Park. If you'd like to summit the highest point in Virginia, turn left at the junction, and you will soon enter a cool, shady, Christmas-tree-scented forest. Examine the next 10 trees you see to determine the mix of red spruce and Fraser fir here—the balance should tip toward Fraser fir at this higher elevation. In fact, Mount Rogers is the only place in Virginia where Fraser fir can be found, and this point marks the northernmost extent of this southern Appalachian endemic. At higher elevations in northern Virginia, balsam fir assumes Fraser fir's role as the high-elevation conifer.

The Mount Rogers summit can be a bit of a letdown if you were expecting spectacular views, but there are more of these to come on your hike down to Grayson Highlands. Instead, enjoy the shade and large rocks that serve nicely as a lunch spot at the summit, which is delineated by two metal markers on adjacent rock outcrops.

Return on the half-mile spur trail to the Appalachian Trail, turning left at the junction to continue north toward Grayson Highlands

When a Grass Bald Isn't

The only grass bald community listed by the Virginia Natural Heritage Program occurs on Whitetop Mountain. Yet your journey from Elk Garden to the Mount Rogers summit on the Appalachian Trail then down into Grayson Highlands takes you through many bucolic, open meadows that certainly appear to be dominated by grasses. Your hike down to Grayson Highlands will be very open compared with the rest of the Appalachian Trail that you've hiked already. Some mistakenly think these areas are grass balds, with their meadowlike openings, grasses, and wildflowers. While we do not completely understand the origins of grass balds in the southern Appalachians, scientists do know that the open spaces around Mount Rogers are of recent origin and thus do not fit the definition of grass bald.

Destructive logging for red spruce (*Picea rubens*) is the reason why you are enjoying these open meadows today. Arrival of the railroads provided access to many areas in the southern Appalachians in the early 1900s. The economy then shifted from small-scale subsistence logging to logging on a large scale, using the destructive methods of the time. Virgin timber was completely removed from the flanks of Mount Rogers because of its proximity to the railroad at Whitetop Station and the railroad's need for profits, with estimated yields of 50,000 to 100,000 board feet per acre. In addition, hot slash fires from logging debris burned deep into these rocky slopes, causing long-term damage. Fires were especially damaging to spruce-fir forests, with their deep, spongy organic soils that you might have noticed

State Park. While most of your hike so far has been a steady climb, the remainder of this hike is a descent. You'll pass the boundary sign again for the Lewis Fork Wilderness and, soon after, come to the well-maintained Thomas Knob shelter at 5,400 ft. If you're curious, check the trail register to see who's been through there recently—the stories and notes can make for fun reading.

You will soon encounter numerous trail junctions in this area. The first, about a mile from the Thomas Knob shelter, is at Rhododendron Gap, about 5,500 ft. Read the signs carefully and look for the rectangular white blaze to stay on the Appalachian Trail. The rhododendron here is the species that favors high elevations, Catawba rhododendron (*Rhododendron catawbiense*), with its abundant magenta blooms in May.

as you climbed to the summit of Mount Rogers. Following this wholesale destruction, the landscape that remained was completely denuded. Recall from your exploration of spruce-fir forests that these species grow in shaded environments with organic soils; following these hot fires, spruce and fir seedlings did not find a hospitable environment, and the forest did not regenerate. As you hike through the open areas, notice the large areas of exposed rocks and thin soils. You might look carefully to see if you can find traces of old stumps as well.

What happened next? Subsequent grazing by domestic livestock kept these treeless slopes and ridgelines open. Herds of cattle and wild ponies help keep large swaths of meadow open today (with the exception of the Lewis Fork Wilderness, which, as a wilderness, cannot be managed). As a result of grazing and soil alteration, trees have never reestablished themselves on the open slopes below the Mount Rogers summit.

However, walking through these rocky, open meadows is a rewarding experience, giving you the feeling of being above treeline. The destruction of the past forest has left evidence right out in the wide-open spaces, but it isn't noticeable to casual visitors, who flock to these slopes for their lovely views and shaggy ponies. Both the US Forest Service and Grayson Highlands State Park actively manage these meadows as open areas, using a combination of prescribed fire and grazing. Just don't call them grass balds—they are a recent human artifact, albeit a scenic one.

The ridgeline you are traveling is Wilburn Ridge. Numerous spur trails climb to rock outcrops; any of these is a great place for a rest stop with spectacular views on a clear day. The Appalachian Trail crosses the boundary into Grayson Highlands State Park through another fence/turnstile, and you'll see a kiosk in front of you with a map of the park trails. Continue to follow the Appalachian Trail blazes into the park, toward Massie Gap. There are several trail options that lead down to Massie Gap; the Appalachian Trail spur trail is the most direct. Follow its blue blazes once you reach and cross a gravel road at about 8.5 mi. You'll then turn right to descend the final distance toward Massie Gap. Your legs may be tired, but take time to enjoy the views as you descend the final switchbacks into Massie Gap.

As you rest your legs and contemplate car-shuttling back to the trailhead, reflect on what you've experienced on this hike. Its length and relatively gradual elevation changes make for a great opportunity to monitor the individual entries and exits of species along the elevation gradient. These make for very gradual transitions among natural communities and confound our attempts to pigeonhole each stand into a particular community type, however attractive that may seem!

As you complete this hike, keep in mind that natural areas are dynamic places. Natural disturbances of various kinds, such as fire, outbreaks of insects or disease, and windthrow, change portions of the landscape. Add to this the often patchy effects of humans and the evolution of land management with timber harvest, fire, domestic animals, and other forms of agriculture, and a very complex landscape pattern emerges. One of the fundamental goals of this book is to help you make some sense of all this complexity. We hope you've enjoyed doing so on today's hike!

30. TWIN PINNACLES TRAIL

Why Are We Here?

It is difficult to choose just one hike from among the excellent offerings at Grayson Highlands State Park; there are many short loop trails that are ideal for morning or afternoon explorations, with something for everyone. If you seek rushing streams, waterfalls, history, wildflowers, or views, all can be found within the park's boundaries.

Your explorations will take you to the high-elevation forests at the northern extent of the area covered by this book. If you are familiar with high-elevation forests further to the south in the southern Appalachians, you'll notice that the distributions of familiar species and natural communities are shifted to somewhat lower elevations at the higher latitude of Grayson Highlands. This is a pattern that will continue far to the north of this book's coverage; indeed, spruce-fir forests can be found all the way down to sea level in Maine and Canada. You will also see the lasting effects of past human activities, and

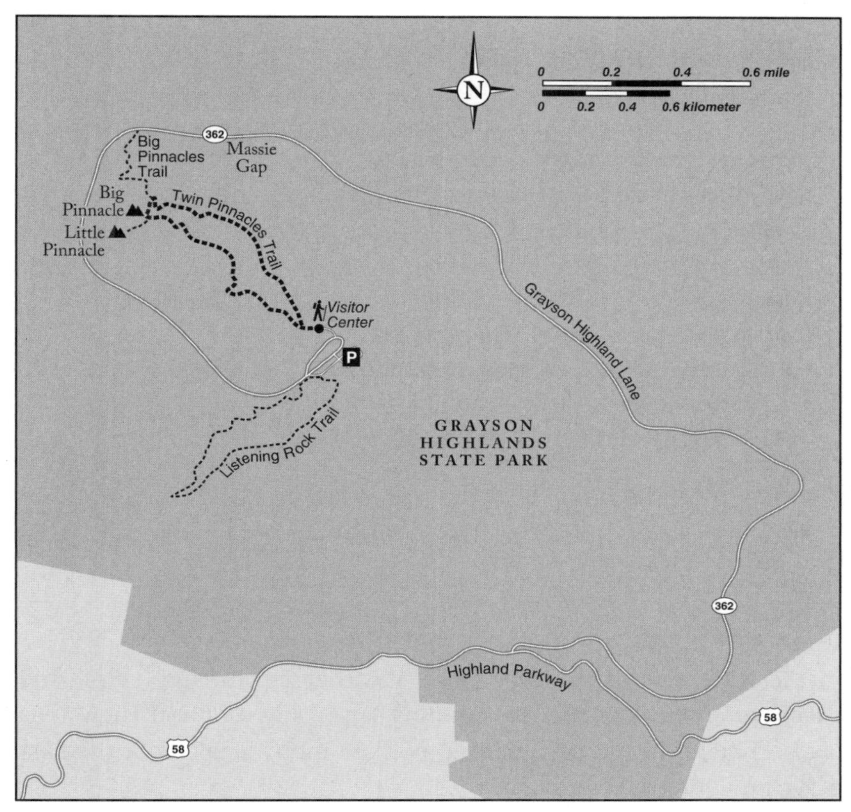

30. Twin Pinnacles Trail

along the way, you will meet two of our most common boreal trees and see some unique rock types found only in this area.

We'll conclude this hike with a sidebar that considers the major influences on the distribution of natural communities in the southern Appalachians. If this is your first hike with us, consider this an introduction to much of what's to come in future hikes. And for the rare individual who has journeyed with us for all 30 hikes, please consider this your capstone lesson!

While you explore our lessons in forest ecology, be sure to take time to experience the sheer joy that comes from a brisk hike in our beautiful southern mountains. Summiting high peaks is something that appeals to nearly everyone, and while the Twin Pinnacles (both peaks on the same Haw Orchard Mountain) are not the highest points

30. TWIN PINNACLES TRAIL

Highlights: Exceptional views from the two highest peaks in Grayson Highlands State Park. From their summits, you can sometimes see wild ponies and enjoy outstanding views of the southwestern Virginia highlands.

Natural Communities: Northern hardwood forest

Elevation: 4,935 ft (trailhead); 5,089 ft (Little Pinnacle, highest point in Grayson Highlands State Park); 5,068 ft (Big Pinnacle)

Distance: 1.6-mi loop, which includes a short spur to the summit of Big Pinnacle

Difficulty: Easy, with a few short, steep climbs

Directions: From Damascus VA, drive US 58 East (or US 58 West from Independence, VA) to the turnoff for Grayson Highlands State Park at VA 362. Drive to the visitor center parking area (small day-use fee). The trailhead is to the left of the visitor center.

GPS: Lat. 36° 37′ 27.9″ N; Long. 81° 29′ 58.7″ W (trailhead)

Information: Grayson Highlands State Park, 829 Grayson Highland Lane, Mouth of Wilson, VA 24363, (276) 579-7092

in the state, they come very close. Weather permitting, you'll find exceptional views of the surrounding landscape. Grayson Highlands is also known for its herd of wild ponies, and this is one of the best hikes on which to see them.

The Hike

Make sure you take time to view the movie and the excellent interpretive exhibits in the visitor center before you begin your hike. There is also an interpretive guide for Twin Pinnacles Trail that you can obtain, and you can then follow the numbered signs as you take your walk. Fill your water bottle and exit the back door of the visitor center; the trailhead is to your left. The trail forks immediately; take the left fork to head toward Little Pinnacle.

Along the first quarter mile of the hike, you can quickly master the dominant tree species. The evergreen tree that is present in increasing numbers as you climb is red spruce (*Picea rubens*). This is a commercially valuable tree that was logged almost completely in the early 1900s from this area. The deciduous tree that co-occurs with

After climbing Little and Big Pinnacles, descend into this typical northern hardwood forest to return to the visitor center. Part of the understory is fern-dominated, while other parts have sedges and wildflowers.

red spruce here, but is even more prevalent, is yellow birch (*Betula alleghaniensis*), distinctive with its light, somewhat shiny, peeling bark. You might notice that yellow birch is dominant along the first part of the trail, at lower elevations, while spruce and other species become more prevalent higher up.

Just past the #4 sign at a quarter mile into your hike, you'll see a large tree on your right with darker bark and long, narrow leaves. This is fire cherry (*Prunus pensylvanica*), and its presence provides clues to some of the human history of the park. After substantial logging in the early 1900s, the debris and slash that were left often ignited, causing very hot fires that burned into the organic soil. The result was nutrient-deficient, mineral soil that would support establishment of few species. Fire cherry is one of the species that thrives in these difficult conditions. If you visit the park in late July or August, you might see the fuchsia spikes of fireweed (*Epilobium angustifolium*), another fire-responsive species.

Because of the significant human influence—the combination of logging plus hot fires, followed by intermittent grazing—the forest

What Does the Mountain Think?

In his 1949 *Sand County Almanac,* Aldo Leopold called on his readers to "think like a mountain," taking the long view of the interconnected natural world. Leopold's specific message was the keystone role of wolves in maintaining the balance of predators and prey, and ultimately the natural vegetation, in a healthy ecosystem. Here we take a somewhat different tack in thinking about the vegetation that clothes the southern Appalachian mountains.

If we were to seek a mountain's guidance in our quest to understand its vegetation, we believe that the mountain would ask us to think in five dimensions. The first dimension is vertical; above all others, elevation is a mountain's defining feature. Elevation controls climate; in the southern Appalachians, increasing elevation is expressed in a climate that becomes increasingly cool and wet, with all the attendant changes in vegetation that derive from such change.

Topography makes the mountain, and thinking in this second dimension helps us understand the effect of topographic variation on the expression of natural communities. Vegetation varies, from the tall forests that grow in the rich, deep, well-watered soils of the protected valley bottoms to the stunted forests and scrub that cling to the rocky, shallow, drought-prone soils of the exposed ridges and summits. Aspect plays an important role in this second dimension, with southern exposures feeling the full force of the sun's energy and its tendency to warm and dry the local environment; northern exposures, in contrast, are sheltered from the sun's rays, and they are thus cooler and moister at any given elevation.

A third dimension involves the very stuff the mountain is made of, its varied geology and the soils derived from them. Rocks rich in dark minerals give rise to soils that are rich and circumneutral (near neutrality, a pH of 7.0), with abundant calcium and other base-forming elements that enhance the growth of many plants; rich coves develop on soils thus defined. Rocks comprised largely of silica and iron yield acidic, nutrient-poor soils that may support entirely different vegetation on sites of identical elevation and topographic position. Returning to our immediate example of mountain coves, we expect to find the acidic variant on such soils.

The fourth dimension is one that most mountains understand exceedingly well, and this dimension is time. Time can be measured on many scales. On an ecological time scale, we observe the effects of catastrophic disturbance and the recovery of natural communities that follows. Fires, windstorms, landslides, floods, and the human interventions of logging and agriculture set the stage for succession, the change in vegetation through time that unfolds after an incident of disturbance. On a longer time scale, we see the effects of changing climate, a process as old as the earth, but one now being accelerated by human activities. In deep time, evolution itself becomes a force defining the very nature of the organisms that inhabit the earth. Time propels the three-dimensional space defined by elevation, topographic position, and geology/soils along a trajectory that imposes kaleidoscopic variation on that space.

The final dimension is one that even the mountain is only dimly aware of, and it towers beyond the mountain itself. In this dimension, we discover Tobler's (2004) first law of geography: "Everything is related to everything else, but near things are more related than distant things." In other words, as we travel from one mountain to another, we find differences in vegetation that are simply the result of our having traveled some distance from our starting point. Such differences may have obvious causes, like the cli-

Hikers of all ages can learn to read the forested landscape! As John Muir (1901) wrote, "Climb the mountains and get their good tidings. Nature's peace will flow into you as sunshine flows into trees."

matic changes we encounter by moving to higher or lower latitudes, but also the more subtle causes associated with geographic isolation.

Working in the context of these five dimensions both opens our eyes and deepens the mystery every time we pause on the trail to pull up a comfortable rock with a sweeping view of the southern Appalachians. Taking time to consider the mountain from such a perch and to ask the mountain "why?" deepens our intimacy with the place and fosters both our understanding and our wonder. In the *Voyage of the Beagle* (1909), Charles Darwin noted, "The pleasure derived from beholding the scenery and general aspect of the various countries we have visited, has decidedly been the most constant and highest source of enjoyment. . . . It more depends on an acquaintance with the individual parts of each view: I am strongly induced to believe that as in Music, the person who understands every note will, if he also has true taste, more thoroughily enjoy the whole; so he who examines each part of [a] fine view may also thoroughily comprehend the full and combined effect. Hence a traveller should be a botanist, for in all views plants form the chief embellishment. Group masses of naked rocks, even in the wildest forms; for a time they may afford a sublime spectacle, but they will soon grow monotonous; paint them with bright and varied colours, they will become fantastick; clothe them with vegetation, they must form, at least a decent, if not a most beautiful picture."

here is difficult to characterize. On the whole, however, we would describe it as a northern hardwood forest dominated by yellow birch. Also, along the Twin Pinnacles loop, you'll discover that wildflowers are not especially rich in the understory, with much of the color show coming from flowering shrubs that thrive in the thin, acidic soil.

As you continue up the trail, you will see other species join the mix, and identifying individual tree species becomes more difficult. Before reaching Little Pinnacle at 0.3 mi, you'll reach a dense, gnarly, short-statured forest consisting mostly of spiny hawthorn (*Crataegus* sp.), giving the mountain its name, Haw Orchard Mountain. Soon you'll emerge from the hawthorn thickets to summit Little Pinnacle at 5,089 ft, the highest point in the park. In May, look for the fiery hues of orange in the flame azalea (*Rhododendron calendulaceum*). In June, you might instead see the magenta blooms of the high-elevation

Catawba rhododendron (*Rhododendron catawbiense*). In early July, anticipate grazing on ripe highbush blueberries (*Vaccinium corymbosum*).

From the Little Pinnacle summit, you can look to the northwest and see Virginia's two highest peaks, Mount Rogers and Whitetop Mountain. After you've enjoyed the views for a bit, find a comfortable place to sit on the exposed rock outcrops, which have a deep purple hue. This rock, called rhyolite, is volcanic in origin, and the Mount Rogers area is the only place in Virginia where it is found.

Descend Little Pinnacle to rejoin the trail, and at 0.4 mi you'll reach the intersection with Big Pinnacle Trail. This is a yellow-blazed spur trail that heads up and to the left to reach Big Pinnacle at 5,068 ft. After enjoying the panorama from the top, which includes a view of Massie Gap down below, head back the way you came along the spur trail, staying on the main trail rather than taking any side spurs (there is a trail leading to Massie Gap from Big Pinnacle, but you will head back and rejoin the Twin Pinnacles Trail). When in doubt, follow the signs toward the visitor center. When you reach the intersection with the Twin Pinnacles Trail, turn left to follow the loop toward the visitor center. You'll soon find yourself back in the yellow-birch-dominated forest dotted with red spruce, and soon the visitor center will be in your sights.

Common Trees and Shrubs of the Southern Appalachians

The Southern Boreal Forest: Spruce-Fir Forest

The evergreen spruce-fir forest is found in the southern Appalachians at elevations above 5,500 ft. Red spruce (*Picea rubens*) (**1**) can be distinguished from Fraser fir (*Abies fraseri*) (**2**) by its rounded, sharp-tipped needles and pendant cones. Fraser fir occupies the highest elevations in pure stands and has flat, soft needles with blunt tips; the cones perch upright on top of the branches like candles. A deciduous tree, mountain ash (*Sorbus americana*) (**3**) has compound leaves that turn brilliant red in the fall, while the smaller shrub hobblebush or witch hobble (*Viburnum lantanoides*) (**4**) is easily identified by its large pairs of veiny, heart-shaped leaves. (Illustrations 1 & 2 by Maryann Roper; 3 & 4 by Dale Morgan.)

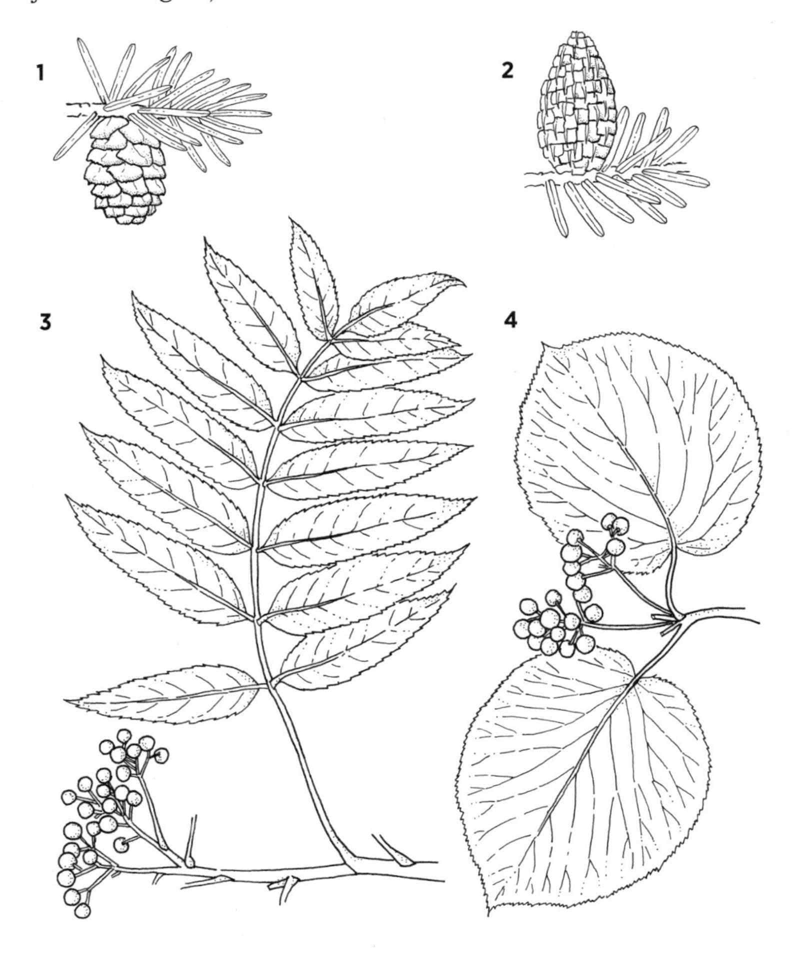

Important Trees of the Northern Hardwood Forest

Northern hardwood forests are dominated by four species: yellow birch (*Betula alleghaniensis*) (**1**), with its shiny bark and prominent lenticels; sugar maple (*Acer saccharum*) (**2**), with its opposite, lobed leaves that resemble the leaf image on the Canadian flag; yellow buckeye (*Aesculus flava*) (**3**), with its opposite, compound leaves that have 5 leaflets radiating from a central point; and American beech (*Fagus grandifolia*) (**4**), with its smooth, gray bark; soft, ovate leaves; and pointed buds. (Illustration 1 by Dale Morgan; 2 & 3 by Maryann Roper; 4 by Ruta Schuller.)

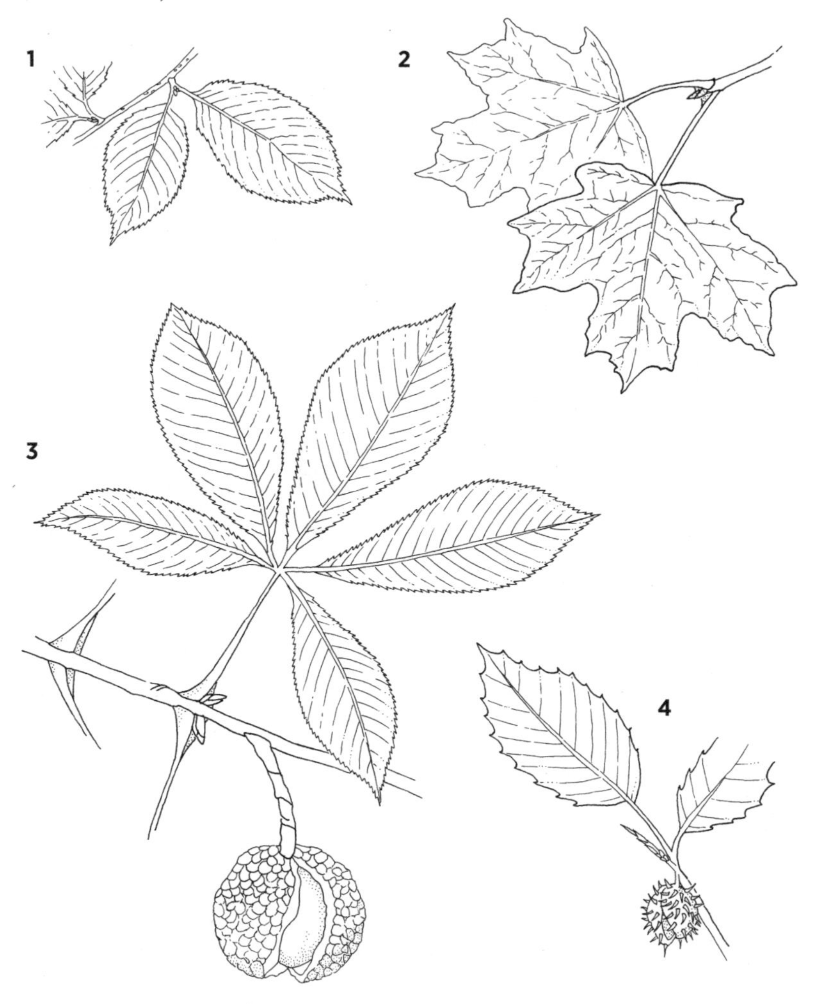

American Chestnut and Oak Family Look-Alikes

All four of these woody plants are in the oak family and have simple, oval-shaped leaves. Chinquapin (*Castanea pumila*) (**1**), with its narrower, densely fuzzy leaves and shrublike form, differs from American chestnut (*C. dentata*) (**2**), which you will likely see as shoots sprouting from old stumps. Both *Castaneas* have sharply serrate leaves. Chestnut oak (*Quercus montana*) (**3**) has rounded, scalloped leaf margins. American beech (*Fagus grandifolia*) (**4**) has smaller and softer leaves than the rest, with less-distinct toothing along the leaf margin. (Illustrations by Ruta Schuller.)

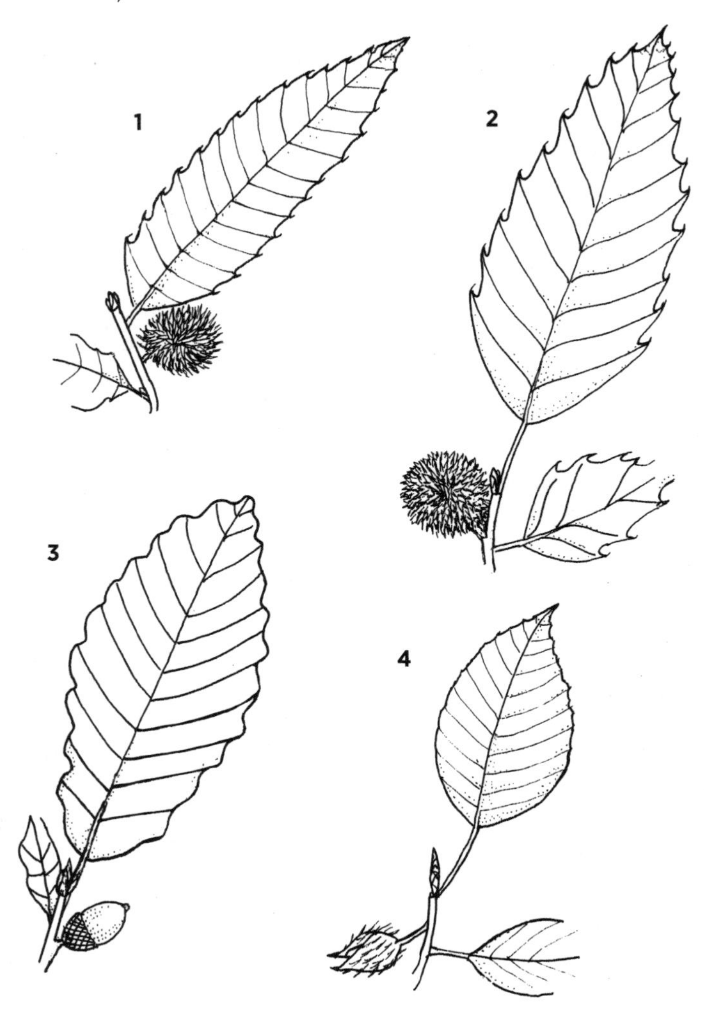

Oaks

White oak (*Quercus alba*) (**1**) is easily distinguished from other oaks by its irregular, rounded lobes, while the sharp, bristle-tipped lobed leaves of northern red oak (*Q. rubra*) (**2**) and black oak (*Q. velutina*) (**3**) are nearly indistinguishable but for the fuzziness on the undersides of black oak leaves, while northern red oak leaves are smooth. Scarlet oak (*Q. coccinea*) (**4**), found most frequently with pine on dry ridges, has deep, C-shaped spaces between lobes, creating a lacy look when seen against the sky. Both white and northern red oaks can be found in a shrubby, gnarled form at higher elevations, while black and scarlet oaks favor lower elevations and dry habitats. (Illustrations by Ruta Schuller.)

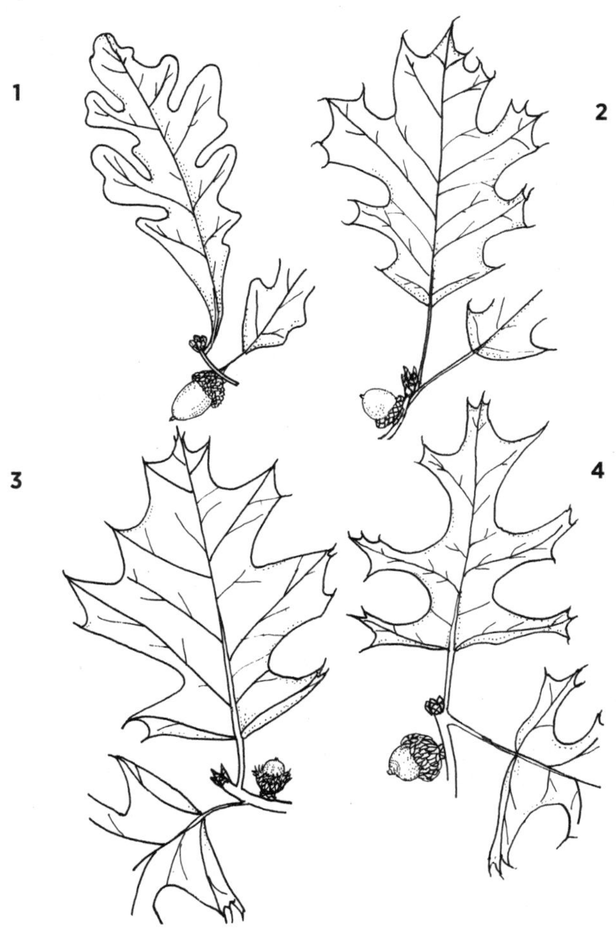

Maples

All maples have leaves with 3 to 5 lobes, positioned oppositely in pairs along the twigs. Red maple (*Acer rubrum*) (**1**) is the most adaptable of the four species, occurring in many habitats and exhibiting a tolerance for all levels of moisture, shade, and disturbance. It has 3 to 5 lobes per leaf and serrate leaf margins. Sugar maple (*A. saccharum*) (**2**) looks most like the leaf pictured on the Canadian flag, with 5 lobes and smooth leaf margins. Striped maple (*A. pensylvanicum*) (**3**) is a small tree with distinctive green stripes on the trunk and 3-lobed leaves with extended tips that help shed water. Finally, mountain maple (*A. spicatum*) (**4**), a small tree, has larger, more veiny leaves than red maple and flowers that grow in upright spikes at the ends of the branches. (Illustrations by Maryann Roper.)

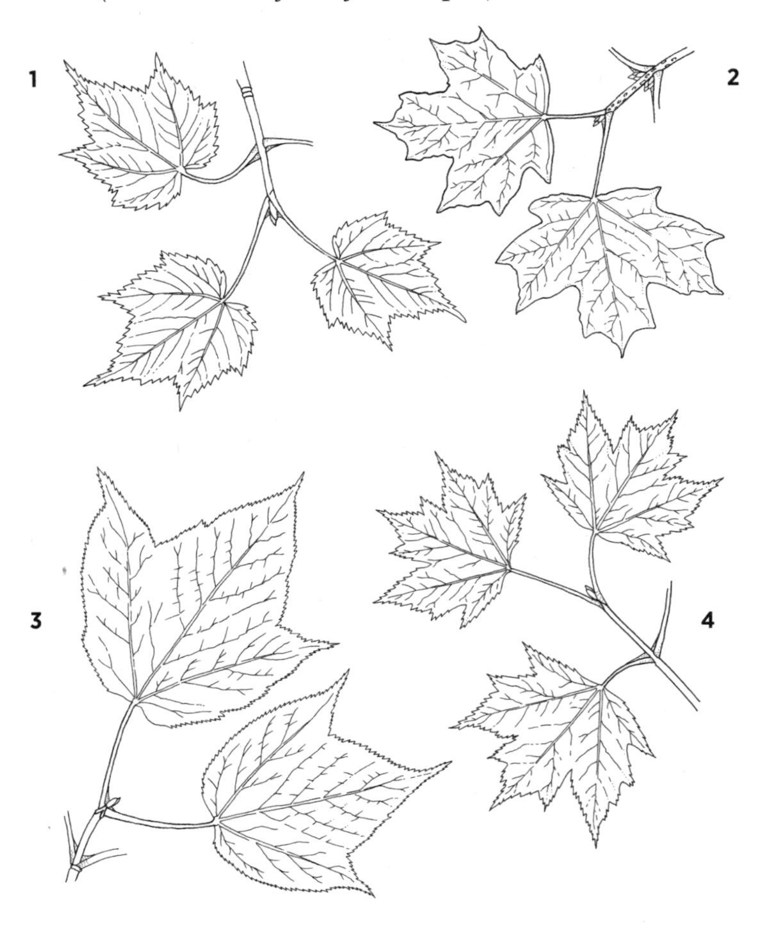

Walnut and Hickories

A tree of lower elevations with nutrient-rich soils, black walnut (*Juglans nigra*) (**1**) is easily distinguished from hickories by its many leaflets and large, green-husked fruits. Pignut hickory (*Carya glabra*) (**2**) has diamond-shaped, braided bark and 5 leaflets with slender stems; the thin-husked nuts sometimes have a protruding "snout" at one end. Shagbark hickory (*C. ovata*) (**3**) also has 5 leaflets per compound leaf; the bark peels away from the main trunk in long, ragged strips, and the nuts are encased in husks greater than one-fourth of an inch thick. Mockernut hickory (*C. tomentosa*) (**4**) has bark similar to that of pignut hickory; mockernut has leaflets in 7s on densely fuzzy, aromatic stems, and its nuts have husks about one-fourth of an inch thick. (Illustrations by Dale Morgan.).

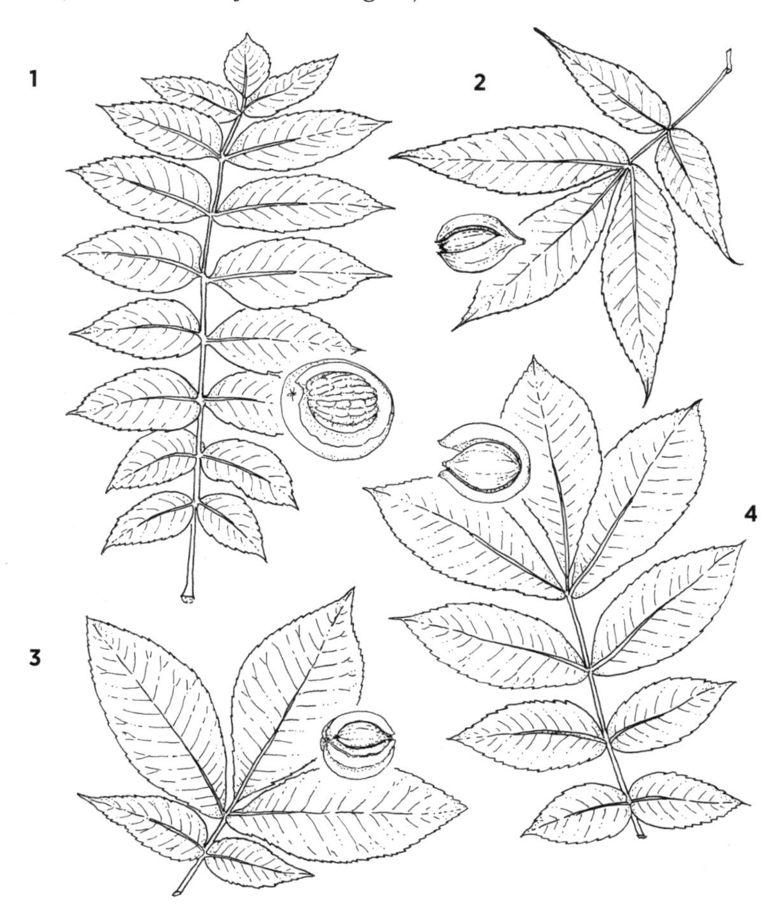

Magnolias

Trees in the magnolia family can be distinguished by thin lines encircling the twigs (these are scars from stipules that fell off) and large, tropical-looking leaves. Cucumber magnolia (*Magnolia acuminata*) (**1**) has large, untoothed, oval leaves and small, oblong fruits that give it its common name. Fraser magnolia (*M. fraseri*) (**2**) is found at higher elevations and is easily distinguished by very large leaves that have ear-lobe-like protrusions at the base. Finally, tulip-tree, or yellow poplar (*Liriodendron tulipifera*) (**3**), dominates cove forests as well as secondary forests below 3,000 ft. Its four-lobed leaves look a bit like cartoon cat faces, with pointed ears at the top and whiskers that flare out below. In mid-spring, you may see on the forest floor the orange and yellow cupped flowers that give tulip-tree its common name. (Illustrations by Maryann Roper.)

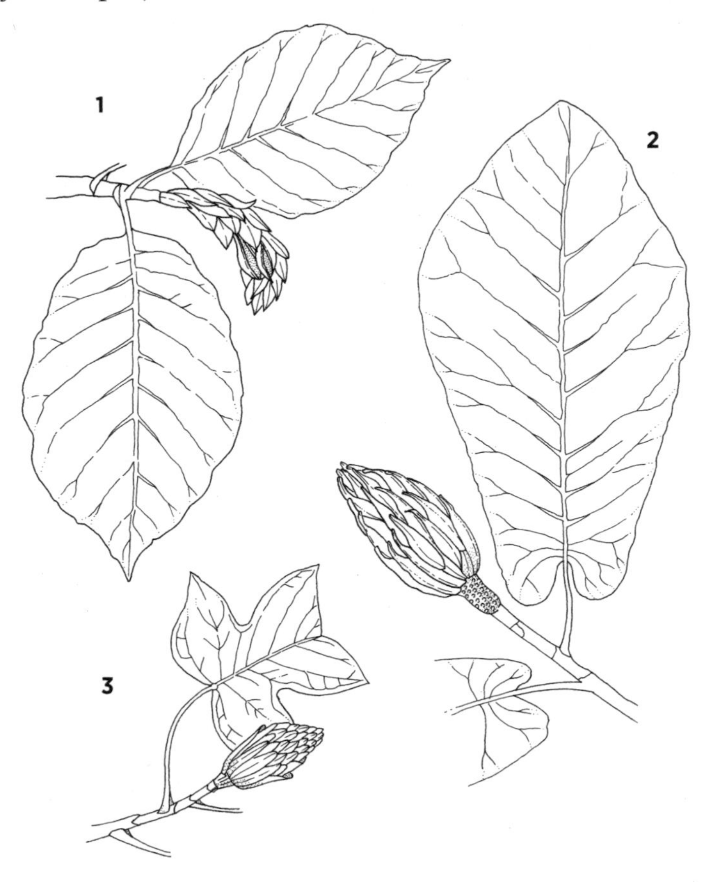

Pines

White pine (*Pinus strobus*) (**1**) is the easiest pine to identify, with 5 needles per bundle and sticky, elongated cones. One flush of branches is produced each year, so these pines are also easy to age. The remaining pines are all yellow pines. Both shortleaf pine (*P. echinata*) (**2**) and Virginia pine (*P. virginiana*) (**3**) can be found at the lowest elevations of the mountains and into the Piedmont. Shortleaf pine usually has 2 to 3 long needles per bundle, and the bark is distinguished by small pits, called resin pockets. Virginia pine, meanwhile, has flaky, reddish bark and short, twisted needles found in bundles of 2. Pitch pine (*P. rigida*) (**4**) can be identified by 3 needles per bundle and, more importantly, tufts of needles growing directly out of the bark; this is a fire adaptation. Lastly, Table Mountain pine (*P. pungens*) (**5**) is found at higher elevations; it has thick needles in bundles of 2, plus heavy, prickly cones attached in whorls along the branches. (Illustrations by Maryann Roper.)

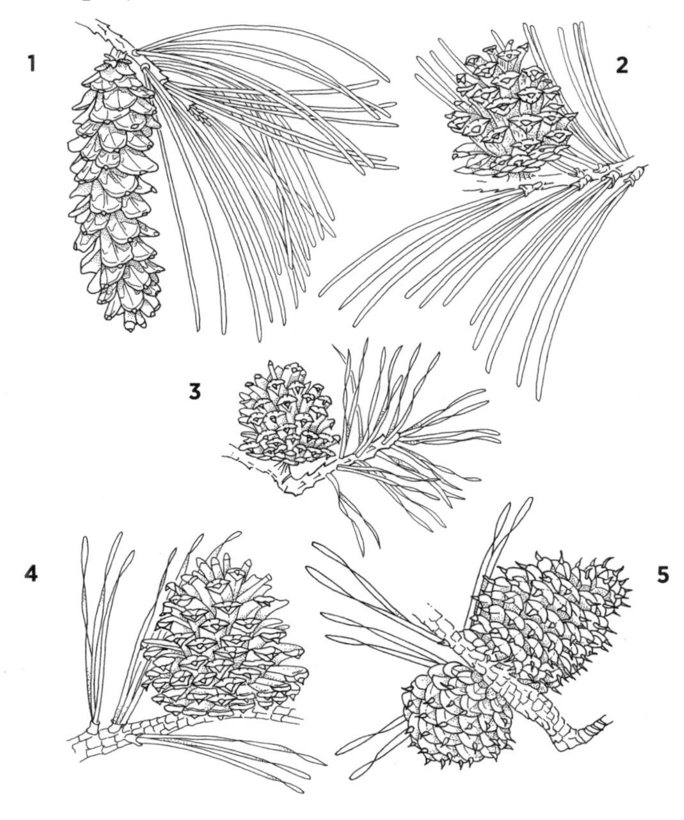

Other Evergreens

Two hemlocks grow in the southern Appalachians. The more common eastern hemlock (*Tsuga canadensis*) (**1**) has flat needles lying in a horizontal plane, while Carolina hemlock (*T. caroliniana*) (**2**) has slightly larger needles that are attached around the entire branch, giving it a bottlebrush appearance, plus larger cones that are more flared. Eastern red cedar (*Juniperus virginiana*) (**3**) grows in a broad variety of habitats and is distinguished by flat, scalelike needles, peeling reddish bark, bushy appearance, and round, waxy blue, fleshy cones. Ground juniper (*J. communis* var. *depressa*) (**4**) also has flattened, scalelike needles, but this small shrub grows prostrate on the ground in exposed rock outcrop habitats. (Illustrations by Maryann Roper.)

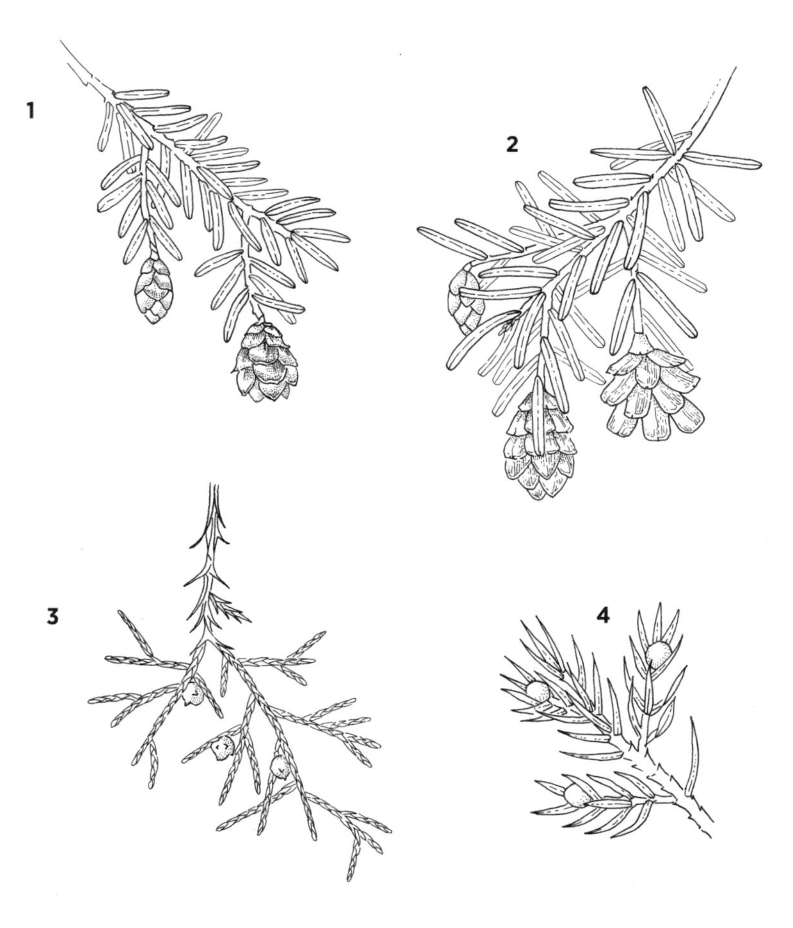

Cherries and Birches

Fire cherry (*Prunus pensylvanica*) (**1**) has long, narrow leaves and smooth shiny bark marked with horizontal lines or lenticels; it is found in secondary forests at higher elevations. Black cherry (*P. serotina*) (**2**) has smaller but broader toothed leaves and grows to canopy size at lower elevations, developing flaky black bark with rounded to oval flakes and horizontal lenticels. Both yellow birch (*B. alleghaniensis*) (**3**) and sweet birch (*B. lenta*) (**4**) have very similar leaves with larger, coarser teeth than those of the cherries, but these birches can be easily distinguished by bark and, to some extent, habitat. Yellow birch has shiny, peeling bark and grows at higher elevations, while sweet birch has darker gray bark with horizontal lenticels adding texture and grows at lower elevations. Young twigs of sweet birch have a pleasant, wintergreen taste when chewed. (Illustrations by Dale Morgan.)

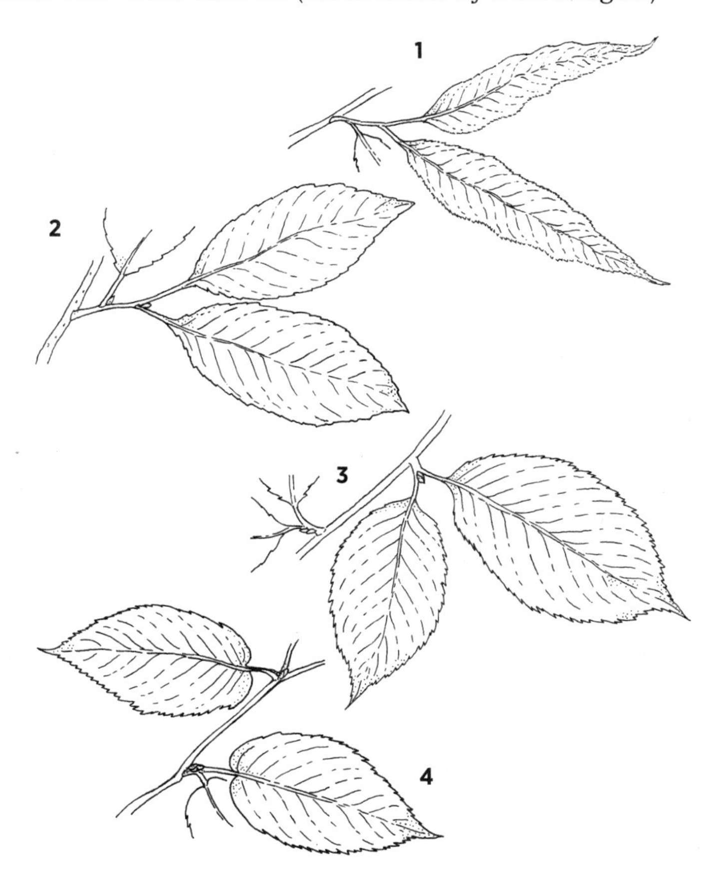

Rhododendrons

Evergreen rhododendrons can be distinguished by ecology and flower color. Catawba rhododendron (*Rhododendron catawbiense*) (**1**) grows at high elevations, often in exposed areas, and can be distinguished by its oval leaves with rounded leaf bases and magenta flowers. Rosebay rhododendron (*R. maximum*) (**2**) is found at lower elevations, often near mountain streams, and its leaves are longer and narrower, with a wedge-shaped leaf base and white to pale pink flowers. Azaleas are deciduous *Rhododendrons*; one distinctive species has bright orange flowers and is thus called flame azalea (*R. calendulaceum*) (**3**). The final example (**4**) could be either Carolina rhododendron (*R. carolinianum*), which is more widely distributed to the north and flowers earlier, or gorge rhododendron (*R. minus*), which is confined to the Blue Ridge Escarpment. Both have smaller leaves than either Catawba or rosebay rhododendron. (Illustrations by Dale Morgan.)

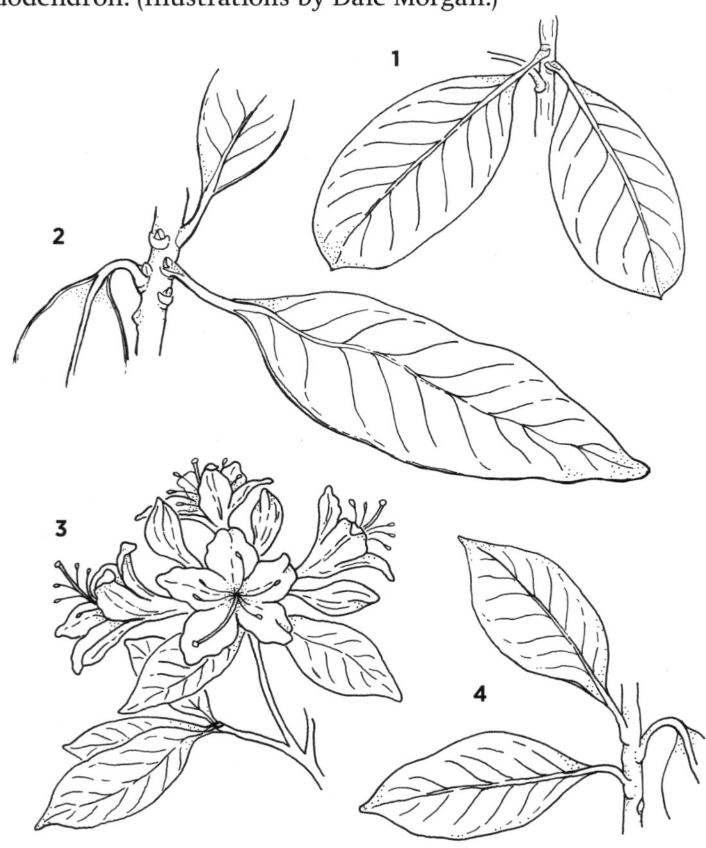

Other Heaths

Sourwood (*Oxydendrum arboreum*) (**1**) is our only tree in the blueberry family, and its sprays of white, urn-shaped flowers make it easy to spot in June. It is an understory tree that prefers dry, acidic soil, and its growth form is most distinctive, as it grows crookedly toward light in the canopy. Growing directly adjacent to mountain streams, doghobble (*Leucothoe fontanesiana*) (**2**) is a low-growing shrub with thick, leathery, saw-toothed leaves. Low-growing sand myrtle (*Kalmia buxifolia*) (**3**) can be found on high-elevation rock outcrops and has tiny, thick leaves. Finally, one of our ubiquitous and best-loved mountain species, mountain laurel (*K. latifolia*) (**4**) is characterized by thick evergreen leaves with a yellow midvein, supported on gnarled, peeling trunks that can sometimes reach the height of a small tree. Sprays of cupped, pale pink flowers open in mid-spring. (Illustrations by Dale Morgan.)

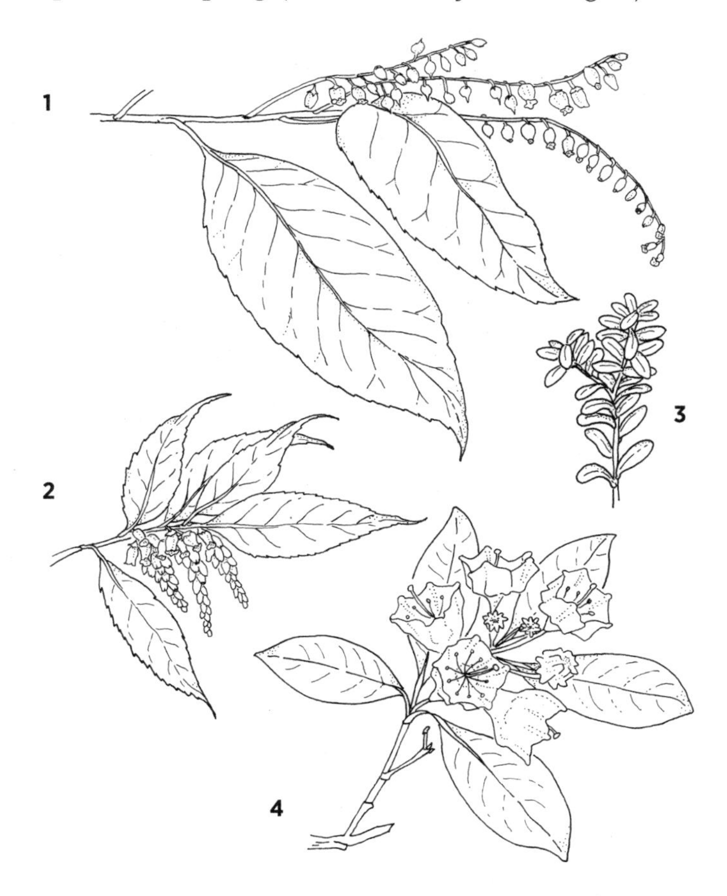

Other Canopy Species

White ash (*Fraxinus americana*) (**1**) has compound, paired leaves and yellowish-gray bark that has a diamond pattern, which varies from open and loose to tight and flat-looking. Black locust (*Robinia pseudoacacia*) (**2**) also has compound leaves, though the leaflets are many, small, and oval. The trees grow in secondary forests and have furrowed, black trunks with spines in pairs at the bases of the leaves. Carolina silverbell (*Halesia tetraptera*) (**3**), a tree found commonly in cove forests, has simple, oval-shaped leaves but an interesting winged fruit and squarish, flaky bark plates that vary from purple to blue-black. Another canopy tree of cove forests, basswood (*Tilia americana*) (**4**) is distinguished by heart-shaped leaves with an asymmetrical leaf base. Finally, black gum (*Nyssa sylvatica*) (**5**) has simple leaves that are widest toward the tip and blocky, corky bark; limbs on younger trees stretch out at a 90° angle from the trunk. (Illustrations by Ruta Schuller.)

Other Understory Species

Many understory trees enrich the diversity of the southern Appalachians. Witch hazel (*Hamamelis virginiana*) (**1**) has yellow, threadlike flowers in late fall and takes the form of a tall shrub or small tree. Its irregular leaf margins and asymmetrical leaf bases make it distinctive. Flowering dogwood (*Cornus florida*) (**2**) has opposite, paired leaves with distinctive arced veins and tight, cobbled bark. The evergreen leaves of American holly (*Ilex opaca*) (**3**) are scalloped and spiny; female trees have bright red berries in the fall. Sassafras (*Sassafras albidum*) (**4**) can have one of three leaf shapes: oval or ovate, mitten-shaped with a single lobe, or tri-lobed. The crushed leaves of this small tree have a spicy scent reminiscent of root beer. (Illustrations by Ruta Schuller.)

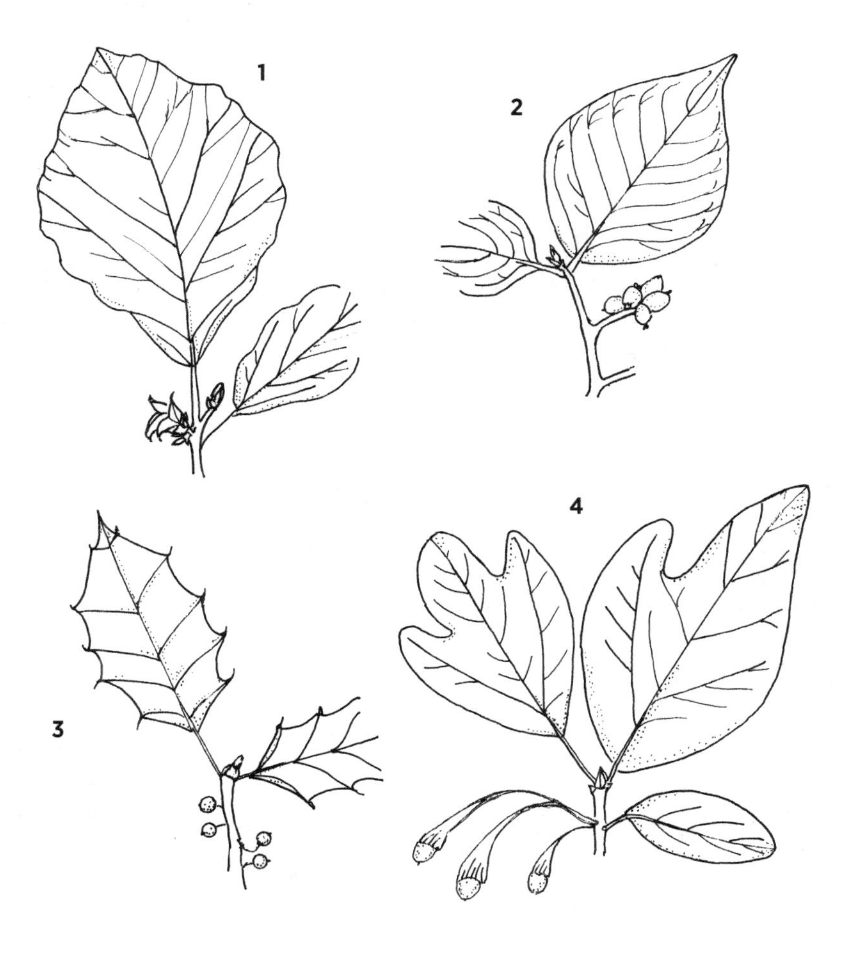

Glossary

Acidic soil: for the purposes of this book, any soil with a pH at or below 5.5
All-aged: a forest stand in which the canopy trees represent a wide range of ages
Alpine: region above treeline in high mountains
Aspect: the compass bearing toward which a sloping surface faces
Boreal: pertaining to high-latitude regions, typically above 45° north latitude; a subarctic region; a forest dominated by spruce and/or fir
Boulderfield: an area largely covered by rocks, typically found in high-elevation ravines in the southern Appalachians
Canopy: the uppermost stratum in a natural community, typically referring to the upper surface of the largest trees in a forest
Circumneutral soil: for the purposes of this book, any soil with a pH at or above 6.0
Codominant: a species sharing dominance in a natural community with at least one other species
Community: the assemblage of plants, animals, and microorganisms that coexist and interact within a particular habitat. See also Natural community.
Continental Drift: See Plate tectonics.
Corridor: narrow area of suitable habitat that connects larger patches of suitable habitat, from the perspective of a particular species or group of species
Cove: a bowl-shaped valley bottom in the southern Appalachians; a canyon
Disjunct: said of a species's range when that range is broken into two discrete regions separated by another region where the species does not occur
Disturbance: any physical disruption to a natural community, whether caused by natural or human influences. Wildfires, rockslides, hurricanes, tornadoes, lightning strikes, agriculture, logging, development, and other processes can all be considered disturbances.
Diversity: See Species richness.
Dominant: a species occupying more space or preempting more resources within its stratum than other species
Ecosystem: the natural community together with its physical environment;

an integrated system of species populations interacting amongst themselves and with their physical environment; the fundamental unit of ecological study

Edge: for the purposes of this book, the boundary between forest and nonforest vegetation

Endemic: said of a species found only in a narrow geographic region

Ericaceous: refers to members of the heath family of plants, the Ericaceae

Escarpment: a steep slope or cliff separating two relatively level landforms

Even-aged: a forest stand in which the canopy trees represent a relatively narrow age class

Exotic invasive species: an exotic species that has reached a sufficient abundance to interfere with the performance of one or more native species

Exotic species: a species that has established itself in a nonnative region

Felsic rocks: those having high levels of iron and silicon and relatively low levels of base-forming elements like calcium and magnesium

Fire suppression: the reduction or elimination of naturally occurring fires from a region by human activities, such as firefighting

Fog drip: atmospheric moisture condensing on needles, leaves, and branches of plants and then dripping to the ground

Forb: any nonwoody plant

Forest floor: the dead and decaying organic debris (leaves, stems, flowers, fruits, and animal remains) found above the mineral soil in a forest

Foundation species: one that plays a central role in determining physical structure or resource availability in a natural community

Fragmentation: the breaking of a contiguous patch of natural community into smaller, unconnected pieces

Gap: a low point along a mountain ridge; an opening in a plant canopy

Herbaceous layer: refers to the ground layer of forbs and tree seedlings beneath the shrub layer of a forest

High-elevation: for the purposes of this book, above approximately 5,000 ft elevation

Hydric: conditions of moisture saturation or flooding, generally unfavorable for plant growth

Hydrophytic: describes plants typically associated with saturated or flooded habitats

Ice Age: any period during the earth's history during which continental ice sheets advance toward the equator from the polar ice packs; specifically, the Pleistocene epoch beginning about 2.6 million years ago (the most recent Pleistocene advance of continental glaciers was at its maximum about 18,000 years before the present and ended about 12,000 years before the present)

Igneous: rock formed by the cooling of magma (molten rock), either beneath or at the earth's surface

Low-elevation: for the purposes of this book, below approximately 3,000 ft elevation

Mafic rocks: those having relatively high levels of iron and base-forming elements like calcium and magnesium

Mesic conditions: of intermediate moisture, generally favorable for plant growth

Mesophytic: describes plants that thrive best in intermediate moisture conditions

Metamorphic: rock that has been physically and/or chemically altered from its original form by intense heating and pressure beneath the earth's surface

Mid-elevation: for the purposes of this book, between approximately 3,000 and 5,000 ft elevation

Native species: any species that is growing in a place where it has not arrived by extraordinary means, typically by human introduction

Natural community: "a distinct and reoccurring assemblage of populations of plants, animals, bacteria, and fungi naturally associated with each other and their physical environment," according to Schafale and Weakley, *Classification of the Natural Communities of North Carolina* (1990)

Niche: describes the full range of environmental conditions tolerated by, and the roles played by, a species within its natural community

Old-growth: a forest stand in which the canopy trees represent a wide range of ages, including some that are very old. Other attributes include the presence of both standing dead and downed tree stems, a height-stratified canopy, and relatively high species diversity.

Parent material: the mineral matter from which a soil is formed

pH: the negative log of hydrogen-ion concentration in water, used as a relative measure of acidity and alkalinity. pH ranges from 0 to 14, with neutrality defined as pH = 7.

Plate tectonics: the movement of the continental plates relative to one another over time

Presettlement: before the arrival of European settlers

Relict: a biological population that remains in a favorable microsite after the environment has become unfavorable in the surrounding landscapes

Richness: See Species richness

Ridge: a narrow, elongated, high-elevation area

Riparian: land area adjacent to a stream or river

Second-growth: a forest that has established itself following a disturbance and has not yet achieved old-growth characteristics

Sedimentary rock: formed from sediments accumulating in the basin of an ocean or large lake, subsequently transformed into rock by intense heating and pressure beneath the earth's surface

Serotinous: describes species like Table Mountain pine (*Pinus pungens*) that retain their seed in unopened fruiting structures and that are stimulated to shed their seed later, perhaps by heat from a passing fire

Shrub layer: refers to the stratum of shrubs and tree saplings beneath the understory of a forest

Soils: the natural bodies covering most of the terrestrial earth, developed from a parent material under the influence of climate, topography, organisms, and time

Species diversity: See Species richness.

Species richness: for the purposes of this book, the number of different, coexisting species in a natural community

Succession: change in the composition and structure of vegetation through time

Tundra: describes treeless communities in environments where the soil stays frozen or partially frozen year-round

Understory: refers to the stratum of tall shrubs and small trees beneath the canopy of a forest

Uneven-aged: See All-aged.

Xeric: conditions of low or limited moisture, generally unfavorable for plant growth

Xerophytic: describes plants that thrive best in low or limited moisture conditions

References and Suggested Readings

References

Bartram, William. *Travels*. Philadelphia: James and Johnson, 1791.

Darwin, Charles. *The Voyage of the Beagle*. New York: P. F. Collier and Son, 1909.

Dugger, Shepherd. *The Balsam Groves of Grandfather Mountain*. Banner Elk, N.C.: Dugger, 1892.

Frost, Robert. "Stopping by Woods on a Snowy Evening." In *New Hampshire: A Poem with Notes and Grace Notes*. New York: Henry Holt and Co., 1923.

Kilmer, Joyce. "Trees." In *Trees and Other Poems*. Garden City, N.Y.: Doubleday and Co., 1914.

Leopold, Aldo. *A Sand County Almanac, and Sketches Here and There*. Oxford: Oxford University Press, 1949.

Mitchell, Elisha. "Personal Journal, 1836." *American Journal of Science* 35 (1839).

Muir, John. *Our National Parks*. Boston: Houghton Mifflin, 1901.

Schafale, Michael P., and Alan S. Weakley. *Classification of the Natural Communities of North Carolina – Third Approximation*. Raleigh: NC Natural Heritage Program, 1990.

The Nature Conservancy. "South Carolina Blue Ridge Escarpment." South Carolina, Places We Protect, 2013. http://www.nature.org/ourinitiatives/regions/northamerica/unitedstates/southcarolina/placesweprotect/southern-blue-ridge-escarpment.xml.

Thoreau, Henry David. *Faith in a Seed: The Dispersion of Seeds and Other Late Natural History Writings*. Washington, D.C.: Island Press, 1993.

Tobler, W. "On the First Law of Geography – A Reply." *Annals of the Association of American Geographers* 94 (2004): 304–10.

Weakley, Alan. *Flora of the Southern and Mid-Atlantic States*. Chapel Hill: University of North Carolina Herbarium, 2011.

Suggested Readings

Adkins, Leonard M. *Hiking and Traveling the Blue Ridge Parkway*. Chapel Hill: University of North Carolina Press, 2013.

Blevins, David, and Michael P. Schafale. *Wild North Carolina: Discovering the Wonders of Our State's Natural Communities*. Chapel Hill: University of North Carolina Press, 2011.

Boyd, Brian. *The Highlands-Cashiers Outdoors Companion*. Clayton, Ga.: Fern Creek Press, 2007.

Burnham, Bill, and Mary Burnham. *Hiking Virginia*. Guilford, Conn.: Globe Pequot Press, 2013.

Clark, John, and John Dantzler. *Hiking South Carolina*. Guilford, Conn.: Globe Pequot Press, 1998.

De Hart, Allen. *North Carolina Hiking Trails: The State's Most Comprehensive Trail Guide*. 4th ed. Boston: Appalachian Mountain Club, 2005.

Edwards, Leslie, Jonathan Ambrose, and L. Katherine Kirkman. *The Natural Communities of Georgia*. Athens: University of Georgia Press, 2013.

Frankenberg, Dirk, ed. *Exploring North Carolina's Natural Areas: Parks, Nature Preserves, and Hiking Trails*. Chapel Hill: University of North Carolina Press, 2000.

Great Smoky Mountains Association. *Hiking Trails of the Smokies*. Gatlinburg, Tenn.: Great Smoky Mountains Association, 2003.

Harper, Francis. *The Travels of William Bartram — Francis Harper's Naturalist Edition*. Athens: University of Georgia Press, 1998.

Homan, Tim. *The Hiking Trails of North Georgia*. Atlanta: Peachtree Publishers, 2001.

Johnson, Randy. *Hiking North Carolina*. Guilford, Conn.: Globe Pequot Press, 1996.

Miller, Joe. *100 Classic Hikes in North Carolina*. Seattle: The Mountaineers Books, 2007.

Pfitzer, Donald W. *Hiking Georgia*. Guilford, Conn.: Globe Pequot Press, 1996.

Roark, Kelly. *Hiking Tennessee*. Guilford, Conn.: Globe Pequot Press, 2009.

Spira, Timothy P. *Wildflowers and Plant Communities of the Southern Appalachian Mountains and Piedmont: A Naturalist's Guide to the Carolinas, Virginia, Tennessee, and Georgia*. Chapel Hill: University of North Carolina Press, 2011.

Stewart, Kevin G., and Mary-Russell Roberson. *Exploring the Geology of the Carolinas: A Field Guide to Favorite Places from Chimney Rock to Charleston*. Chapel Hill: University of North Carolina Press, 2007.

Wells, B. W. *The Natural Gardens of North Carolina*. Chapel Hill: University of North Carolina Press, 2002.

White, Peter, Tom Condon, Jane Rock, Carol Ann McCormick, Pat Beaty, and Keith Langdon. *Wildflowers of the Smokies*. Gatlinburg, Tenn.: Great Smoky Mountains Natural History Association, 2003.

Index

Steph Jeffries is a naturalist at heart and a forest ecologist by training. Her fascination with the mountains began at the end of a rope, rock climbing in Pisgah National Forest and Linville Gorge while a student earning her B.S. in marine science at the University of South Carolina. After earning a Ph.D. in forestry from North Carolina State University, Steph taught in a variety of university, public, and outdoor settings before returning to NC State as a faculty member in 2011. She also loves teaching at the Highlands Biological Station and the NC Botanical Garden. Outdoors, she shares her passion for ecology and for the natural world with people of all ages – her two sons most especially. She is also an avid runner and has published work in Trail Runner magazine, among other venues. To keep up with Steph, visit her blog, Running with Scissors (stephjeffries.wordpress.com).

Growing up in New England, **Tom Wentworth** delighted in exploring the mountains, rivers, lakes, and coastal regions of Maine and New Hampshire. After earning his B.A. in biology at Dartmouth College and his Ph.D. in plant ecology at Cornell University, Tom moved to North Carolina, where he has spent nearly four decades on the faculty at North Carolina State University in Raleigh. Tom's earliest teaching and research ventures led him to the southern Appalachian Mountains, which in many ways resemble the northern mountains he enjoyed as a youth. With the guidance of outstanding mentors, Tom became familiar with the diverse flora and natural communities of the region, especially in Great Smoky Mountains National Park and the national forests surrounding Highlands, North Carolina. Tom has shared his love and knowledge of the flora and vegetation of the southern Appalachians with hundreds of students, through field trips in his plant ecology course at NC State and in courses offered through the Highlands Biological Station.